2021 **RMP** UPDATE

A Report of the Regional Monitoring Program for Water Quality in San Francisco Bay

NOTE TO READERS: The RMP produces two types of summary reports: *The Pulse of the Bay* and the *RMP Update*. *The Pulse* focuses on Bay water quality and summarizes information from all sources. The *RMP Update* has a narrower and specific focus on highlights of RMP activities. The next *Pulse of the Bay* will be published in 2022.

DIGITAL VERSIONS of all RMP Updates are available at: www.sfei.org/rmp/update

DIGITAL VERSIONS of all Pulses are available at: www.sfei.org/rmp/pulse

COMMENTS OR QUESTIONS regarding the RMP Update can be addressed to Dr. Jay Davis, RMP Lead Scientist, (510) 746-7368, jay@sfei.org.

SUGGESTED CITATION: San Francisco Estuary Institute (SFEI). 2021. RMP Update 2021. SFEI Contribution #1057. San Francisco Estuary Institute, Richmond, CA.

VERSION NUMBER: 1.1 (12/01/21)

To download this report please visit www.sfei.org/rmp/update

PREFACE

The overarching goal of the Regional Monitoring Program for Water Quality in San Francisco Bay (RMP) is to answer the highest priority scientific questions faced by managers of Bay water quality.

The RMP is an innovative collaboration between the San Francisco Bay Regional Water Quality Control Board, the regulated discharger community, the San Francisco Estuary Institute, and many other scientists and interested parties.

The purpose of this document is to provide a concise overview of recent RMP activities and findings, and a look ahead to significant products anticipated in the next two years.

The report includes:

- a brief summary of some of the most noteworthy findings of this multifaceted Program;

- a description of the management context that guides the Program; and

- a summary of progress to date and future plans for addressing priority water quality topics.

Program Highlights — 1

The RMP Top 10 — 2

Coming Attractions — 8

Program Oversight — 10

Program Management — 11

Featured Project:
RMP Monitoring of Contaminants
in San Francisco Bay Fish: 2019 — 12

Recent Publications — 26

Program Impact — 30

The Impact of the RMP on Management Decisions — 32

Regulatory Policies Informed by the RMP — 34

RMP Impact Summaries — 36

Municipal Wastewater — 36

Municipal Stormwater — 37

Industrial Wastewater — 38

Dredgers — 39

Program Area Updates — 40

Status and Trends Monitoring — 42

Emerging Contaminants — 44

Small Tributary Loading — 46

Nutrients — 48

PCBs — 50

Microplastics — 52

Sediment — 54

Acknowledgements — 56

CONTENTS

PROGRAM
HIGHLIGHTS

THE RMP TOP 10

Recent Activities and Accomplishments

1 General
National Environmental Achievement Award for the RMP

The National Association of Clean Water Agencies (NACWA), which represents public clean water utilities in the US, has given the RMP a 2021 Watershed Collaboration Award. NACWA's National Environmental Achievement Award Program recognizes NACWA member agencies that have made outstanding contributions to environmental protection and the clean water community. The Watershed Collaboration award is presented for an outstanding watershed-based collaborative management initiative or program focused on cost-effective solutions to environmental challenges. East Bay Municipal Utility District, one of the RMP's municipal wastewater participants, submitted the application and received the award on behalf of the RMP.

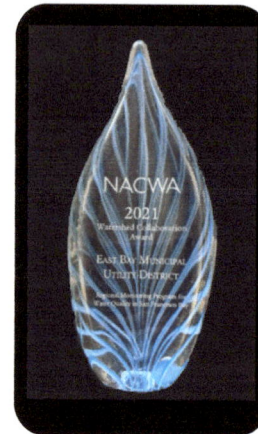

According to the statement accompanying the award, "the Environmental Achievement Award recognizes how the RMP's collaborative approach benefits the environment by defining, investigating, and providing the critical information decision-makers need to proactively protect water quality in San Francisco Bay; benefits dischargers by generating robust and accurate data; and benefits the community by utilizing cost-effective methods in protecting the environment while minimizing the financial burden placed on rate payers. The RMP is an exceptional example of a regional collaboration between many and diverse stakeholders including regulators, scientists, dischargers, and non-governmental organizations, all with a common goal of protecting San Francisco Bay. For over 25 years, through sound science and technological evolution, the RMP has informed decision-making for how to best manage a critically important ecosystem. Detailed investigations have been performed for a range of contaminants and environmental issues, including the recent remarkable investigations into nutrients, microplastics, and PFAS. RMP is truly one-of-a-kind and serves as a model of collaboration nationwide."

The RMP is an exceptional example of a regional collaboration between many and diverse stakeholders including regulators, scientists, dischargers, and non-governmental organizations, all with a common goal of protecting San Francisco Bay

MORE INFORMATION

NACWA WEBSITE: https://www.nacwa.org/about-us/awards/national-environmental-achievement-award-program/neaa-2021-honorees

2 CECs
The Tire Preservative 6PPD

A collaborative, RMP-supported study of contaminants of emerging concern in Bay Area stormwater found a highly toxic tire-related contaminant at levels that are lethal to coho salmon at four sites. The contaminant, derived from a tire preservative, was recently discovered to be responsible for high levels of coho salmon mortality in Puget Sound streams.

The research team, led by scientists at the University of Washington and Washington State University, published the results of their investigation in the journal Science. The chemical, 6PPD-quinone (6PPDQ), a transformation product of the antioxidant chemical 6PPD, can wash into streams along with tire wear particles when it rains. RMP scientists collected samples from nine Bay Area streams and storm drains during storm events; four contained levels of 6PPDQ above the concentration at which half the coho salmon die after a few hours of exposure in laboratory experiments. A previous study by SFEI found that nearly half of the estimated seven trillion microplastic particles in urban stormwater flowing through local streams into the Bay could potentially be linked to tire wear. These new findings indicate such particles can be toxicologically relevant.

Coho salmon no longer reside in San Francisco Bay and its streams, but they are being restored to coastal streams from Santa Cruz to Sonoma County. Researchers are also concerned that steelhead trout and Chinook salmon exhibit some sensitivity to tire rubber chemicals, and studies are ongoing on those species.

MORE INFORMATION

SFEI WEB PAGE, INCLUDING LINKS TO MEDIA COVERAGE: https://www.sfei.org/news/toxic-tire-contaminant-found-bay-area-stormwater

SCIENCE ARTICLE: A ubiquitous tire rubber-derived chemical induces acute mortality in coho salmon. Tian, Z.; Zhao, H.; Peter, K.T.; Gonzalez, M.; Wetzel, J.; Wu, C.; Hu, X.; Prat, J.; et al. 2020. Science 371 (6525), 185-189. https://www.science.org/doi/10.1126/science.abd6951

Informing Regulatory Decisions: DTSC Evaluation of Chemicals in Motor Vehicle Tires

Recent studies by the San Francisco Estuary Institute (SFEI) and the RMP have pinpointed chemical and microplastic contamination from tires as a rising concern in the Bay. Tire ingredients of interest include zinc, which has many documented adverse health effects in aquatic life, 6PPD, and other chemicals that may also pose concerns to aquatic life.

The California Department of Toxic Substances Control (DTSC) has begun a process to evaluate regulating the presence of zinc and 6PPD in tires under the Safer Consumer Products Program. Specifically, DTSC's 2021-2023 Priority Product Work Plan added motor vehicle tires as a new product category, in part due to concern about the potential for adverse impacts to aquatic organisms from exposure to chemicals in these products. DTSC is also considering several other chemicals found in motor vehicle tires and in the Bay, including benzothiazoles, chlorinated paraffins, 1,3-diphenylguanidine, (methoxymethyl) melamines, octylphenol ethoxylates, and polycyclic aromatic hydrocarbons (PAHs).

In July 2021, DTSC held a two-day virtual public workshop to discuss this issue, with SFEI scientists providing scientific input to inform decision-making, alongside international scientists, industry experts, and impacted stakeholders. Specifically, Dr. Kelly Moran provided broad scientific background on tires as a source of contaminants via a presentation titled, "How Tire Particles and Chemicals Reach California's Aquatic Environments." Later, Dr. Rebecca Sutton documented the presence of a broad range of tire ingredients in San Francisco Bay, and updated the agency on future monitoring efforts.

Should DTSC decide to regulate ingredients in tires, manufacturers would begin a process designed to identify more environmentally safe alternatives while maintaining tire safety and performance standards. Workshop materials and agendas can be found on the DTSC website.

MORE INFORMATION

SFEI WEB PAGE: https://www.sfei.org/news/sfei-experts-assist-california%E2%80%99s-safer-consumer-products-program

DTSC WEB PAGE: https://dtsc.ca.gov/scp/chemicals-in-motor-vehicle-tires/

3 Microplastics/Small Tributaries
Conceptual Models of Microplastics in Urban Stormwater

The major field study of microplastics in the Bay completed in 2019 identified stormwater to be a dominant pathway. Given this finding, the RMP Microplastics Workgroup prioritized the development of stormwater conceptual models for microplastics. A robust conceptual understanding will help inform management actions to address microplastic pollution and identify data gaps related to Workgroup management questions. Development of the conceptual models was also a priority for the California Ocean Protection Council, which provided most of the funding for the work, to inform development and implementation of a Statewide Microplastics Strategy related to microplastic materials that pose an emerging concern for the ocean.

The Bay microplastics study indicated a few likely sources for some of the most abundant types of microplastics in urban runoff that were prioritized for conceptual model development. Specifically, conceptual models were developed for cigarette butts and associated cellulose acetate fibers, fibers other than cellulose acetate, single-use plastic foodware and related microplastics, and tire particles. Conceptual models for each of these particle types are valuable tools to refine source identification and elucidate potential source-specific data gaps and management options.

The report resulting from this project will be published in the fall of 2021. For the first time, this report identifies major sources of microplastics in urban runoff, provides a broad menu of potential management strategies for California, and highlights priority data needs to inform California's future management decisions. A few highlights of the information presented include an estimate of 3-5.5 kg/capita/year of tire wear emissions in the US, discussion of the two particle types associated with toxicity to aquatic organisms (tires and fibers), identification of specific sources (e.g., clothing dryers), and a review of mitigation options (e.g., tire wear collection on vehicles).

MORE INFORMATION

SFEI WEBSITE: https://www.sfei.org/projects/microplastics

4 Small Tributaries
Dynamic Modeling of Watershed Loading

In 2020, development began on a tool that will be extremely valuable in assessing and managing inputs of contaminants and sediment from watershed runoff: a new regional watershed dynamic model (WDM).

The WDM is classified as a dynamic model because it is based on a quantitative representation of hydrological and pollutant transport processes. It will provide updated estimates of contaminant concentrations and loads from all local watersheds across various land use, land cover, soil, topographic, and geological conditions. The WDM is an enhanced version of a previous dynamic model for copper load estimation for the Bay Area, with higher spatial resolution and refined process representations. The potential impacts of projected changes in weather patterns and land use, various control measures, or other future scenarios on contaminant loads and trends can then be explored.

The model will initially be used to evaluate PCB and mercury loadings at watershed and regional scales. The trial using these two well-sampled contaminants will provide a proof of concept and basis for modeling other constituents, such as contaminants of emerging concern (CECs), sediment, and nutrients. The first step in the multi-year project was to develop the hydrologic model. A report on this work that documents model development and hydrologic calibration was published in April 2021. The completed hydrologic model is performing well, reproducing the timing and peaks of runoff events as well as the annual and intra-annual variation of hydrological processes. The hydrologic model setup and calibration serves as a solid foundation for future model developments on sediment and contaminant simulations. The focus in 2021 is on developing and calibrating the sediment model. The contaminant models will be developed in 2022 and 2023.

MORE INFORMATION

RMP TECHNICAL REPORT

San Francisco Bay Regional Watershed Modeling Progress Report, Phase 1. Zi, T.; McKee, L.; Yee, D.; Foley, M. 2021. SFEI Contribution # 1038. San Francisco Estuary Institute: Richmond, CA. https://www.sfei.org/documents/san-francisco-bay-regional-watershed-modeling-progress-report-phase-1

5 Small Tributaries
Advanced Data Analysis

Over the past seven years, the RMP, in collaboration with county stormwater programs, has generated an extensive dataset on PCBs in stormwater. Reconnaissance sampling (a screening approach based on collection of a composite water sample during a single storm) has been conducted in 91 watersheds, and some watersheds have been identified as higher priorities for management attention based on elevated PCB concentrations on exported particles. Some watersheds with more moderate concentrations, however, might still have smaller patches of contaminated source areas that merit management attention.

To make better use of data from watersheds exhibiting moderate or lower PCB concentrations, in 2018 the RMP began exploring new data analysis methods based on loads and yields and PCB congener patterns to provide additional information to support management decisions. A pilot study evaluating data from watersheds in San Mateo and Santa Clara Counties resulted in two new analysis methodologies: one based on a deeper analysis of PCB loads and yields (loads per unit area) that takes variation in storm size and land use composition into account; and one based on examining variation in PCB congener patterns, which can help to pinpoint important source areas.

A final report that is planned for publication in late 2021 will describe enhancements to the loads and yields method and then apply both methods to a much larger data set: approximately 130 watersheds for the loads and yields method and 75 watersheds for the PCB congener method. This analysis revealed some new insights on areas within watersheds to consider for PCB management and also provided corroborating evidence for other watersheds where management is already underway.

MORE INFORMATION

RMP TECHNICAL REPORT

Small Tributaries Pollutants of Concern Reconnaissance Monitoring: Application of loads and yields-based and congener-based prioritization methodologies. McKee, L.J., Gilbreath, A.N. In preparation. SFEI Contribution # 1053.

6 Sediment
Updated Assessment of Bay Bathymetry

Mapping the bathymetry of the Bay provides information that is essential for understanding the long-term Bay sediment budget. Bathymetric information can be used to assess how the Bay has responded to changes in sediment supply from the Delta and local tributaries over the decades and to support decision-making on a variety of issues including wetland restoration, strategies for beneficial use of dredged material, and management of legacy contaminants.

In 2014 and 2015 the Ocean Protection Council funded bathymetric surveys of large portions of San Francisco Bay. These data, along with recent smaller-scale surveys by NOAA, USGS, and California State University Monterey Bay were combined to create a revised bathymetric digital elevation model (DEM) of the whole Bay (Lower South Bay, South Bay, Central Bay, San Pablo Bay, and Suisun Bay). Analysis of these surveys and comparison with the USGS DEMs of earlier surveys has provided an update on the quantities and patterns of erosion and accretion in the Bay over the past 25 to 35 years. The analysis showed the Bay as a whole continued to be erosional in this period, with a net loss of approximately 50 million cubic meters of sediment from 1980s to 2010s. San Pablo Bay and South Bay lost sediment, while Central Bay showed a slight gain. A partial mapping of Suisun Bay indicates a loss of sediment.

MORE INFORMATION

TECHNICAL REPORT AND DATA

USGS DATA RELEASE FOR THE DIGITAL ELEVATION MODEL: https://www.sciencebase.gov/catalog/item/5e0592d6e4b0b207aa094f2a

USGS OPEN-FILE REPORT: coming in October

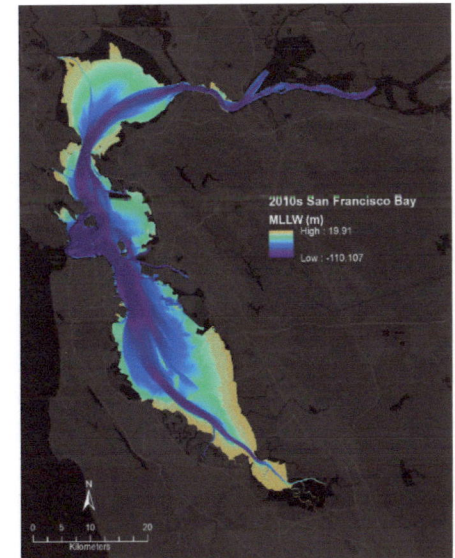

2010s San Francisco Bay MLLW (m)
High : 19.91
Low : -110.107

7 Sediment
Sediment for Survival Report

With partial funding from the RMP, SFEI worked with local, state, and federal scientists to develop the Sediment for Survival report. The report provides a regional sediment strategy that examines the future of sediment in the Bay and informing sediment management for the resilience of tidal marshes and tidal flats to climate change. The report analyzes current data and climate projections to determine how much natural sediment may be available for tidal marshes and tidal flats and how much supplemental sediment may be needed under different future scenarios. These sediment supply and demand estimates are combined with scientific knowledge of natural physical and biological processes to offer multi-benefit strategies for sediment delivery. The analysis and options presented in this report provide land managers with the information needed to make investment decisions related to sustaining wetlands in the face of sea level rise and other climate change stressors.

The report presented three key findings.

1. Tidal marshes and mudflats are unlikely to receive enough sediment naturally to survive sea-level rise this century.

2. Other local sediment sources offer the potential to help maintain tidal marshes and tidal flats that will be resilient as the climate continues to change.

3. Management practices need to change quickly to access these other sources of sediment that can help increase the future resilience of tidal marshes and mudflats.

MORE INFORMATION

REPORT AND MEDIA COVERAGE: SFEI website with links to the report and media coverage:
https://www.sfei.org/news/sediment-survival-report-released

8 Nutrients
Nutrient Moored Sensor Network Findings

Bay Area municipal wastewater dischargers are the principal funders of the Nutrient Management Strategy, a multi-year, multi-million dollar effort to develop the science needed to support management of nutrients in the Bay. The RMP, in a partnership with the NMS, contributes a substantial amount of funding toward the NMS and Bay nutrient studies. Specifically, funds from the RMP support the NMS Moored Sensor Program (supplementing NMS funding of this element) and the USGS water quality cruises (see Item 9).

The Moored Sensor Program was launched in July 2013 to collect nutrient-related water quality data at high temporal frequency as a complement to long-term ship-based monitoring to enhance the characterization of water quality, provide additional mechanistic insights into physical and biogeochemical dynamics, and allow for improved calibration of biogeochemical models.

The sensors provide continuous data on salinity, temperature, chlorophyll, dissolved oxygen, dissolved organic matter, and turbidity at nine stations in South Bay and Lower South Bay. Two stations also have nitrate sensors.

These data show that elevated phytoplankton biomass and low dissolved oxygen are frequently observed in Lower South Bay margin habitats, and suggest that water from the salt ponds introduces high phytoplankton biomass into Lower South Bay sloughs increasing the potential for low dissolved oxygen events. In the summer of 2020, unprecedented smoke from wildfires led to the lowest dissolved oxygen concentrations ever observed by the NMS in the Lower South Bay. The absence of light resulted in a shift in the metabolic balance of the system, causing oxygen concentrations to plummet, putting fish and other biota at risk.

MORE INFORMATION

Nutrient Moored Sensor Program: Program Update. Winchell, T.; Sylvester, Z.; King, E.; MacVean, L.; Trowbridge, P.; Senn, D. 2018. SFEI Contribution # 930. San Francisco Estuary Institute: Richmond, CA. https://sfbaynutrients.sfei.org/sites/default/files/2019_moored_sensor_program_update.pdf

9 Nutrients
USGS Sustains Support for Collaborative Long-term Monitoring

The long-term monitoring program established by the USGS in 1969 to monitor nutrients and primary productivity is one of the longest observational records in a US estuary, and has been instrumental in advancing understanding of primary productivity and nutrient concentrations in the Bay. The RMP began contributing funds to this monitoring as soon as the RMP began in 1993, and in the past decade has supported additional monitoring to expand our understanding of how the Bay is responding to increasing nutrient loading. The Bay has some of the highest concentrations of nitrogen and phosphorus of any estuary in the world, and yet it has not had the same problems with eutrophication (i.e., excessive production of phytoplankton) that many estuaries, including those with lower nutrient concentrations, have suffered.

The USGS long-term water quality monitoring program, led by Dr. Jim Cloern, continued largely unimpeded until 2019 when Cloern retired and the USGS decided not to hire a new scientist to fill his role. Due to the uncertainty around the future of the USGS Bay nutrient cruises, the RMP and USGS have been confined to single year funding agreements with no guarantees of future work. Over the last year, however, the USGS has committed to continuing the long-term monitoring program and providing support for the R/V Peterson and crew. Beginning in October 2021, a multi-year funding agreement between the RMP and USGS California Water Science Center will be in place, allowing the collaborative partnership between the USGS and RMP to continue in support of Bay nutrient monitoring and management.

MORE INFORMATION
USGS WEBSITE: https://www.usgs.gov/mission-areas/water-resources/science/water-quality-san-francisco-bay-research-and-monitoring

10 Status and Trends
Updated Monitoring Design

Status and trends monitoring in the RMP has consisted of the long-term measurement of contaminants in water, sediment, sport fish, bivalves, and bird eggs since the Program began in 1993. In 2002 the RMP began to implement a new sampling design for water and sediment monitoring, and shortly thereafter also evaluated and modified the design for biota monitoring.

After 20 years of implementing this basic design, the Program is conducting a reevaluation of RMP S&T monitoring to ensure that it is optimized to cost-effectively provide the information most needed by Bay water quality managers. The key aspect of the reevaluation is a shift toward a primary focus on contaminants of emerging concern (CECs).

The S&T review began in 2020 and continued through 2021. The review is being conducted by a S&T Workgroup that includes eight external science advisors with extensive expertise in long-term monitoring programs, CECs and legacy contaminants, and statistical analysis. The advisors are working in collaboration with RMP staff and stakeholders to review the existing Program, perform statistical analyses, and define sampling priorities to inform the updated design. The water design was finalized in January; the sediment design in June; and the biota design in August. A synthesis meeting in September ensured that all three designs are well aligned, are providing the necessary data to inform management decisions, and can be completed with available funding.

MORE INFORMATION
SFEI WEBSITE: Forthcoming report on the Status and Trends redesign

COMING ATTRACTIONS

1 **CECs**
CECs in Stormwater

A major, multi-year study to measure CECs in urban stormwater began in 2019 and is continuing through 2022. A long list of CECs is being analyzed, including PFAS, ethoxylated surfactants, phosphate flame retardants, and roadway contaminants. Results will be reported in 2023.

2 **CECs**
QACs in Wastewater

Due to the COVID-19 pandemic and the major use of quaternary ammonium compounds (QACs) as antimicrobial active ingredients, use of these compounds likely increased significantly in 2020. As a result, the RMP launched a special study analyzing QACs in wastewater as well as stormwater and sediment. A report on the results will be available in 2022.

3 **CECs and PCBs**
In-Bay Fate Modeling
Strategy for PCBs and CECs

This study is developing a strategy and multi-year workplan for modeling PCBs and CECs in the Bay. Modeling is needed to address several management questions that are a priority for PCBs, and the platform developed for PCBs will also be applied to answering management questions for CECs and other contaminants. A report will be available in early 2022.

4 **Status and Trends**
Updated Status
and Trends Design

The first major re-design of RMP Status and Trends monitoring since 2002 is underway, with input from external advisors and stakeholders. The primary goal is to optimize the design for monitoring CECs. Completion of the design process is anticipated in early 2022.

5 **Status and Trends**
North Bay Margins Sediment

In 2020, the RMP finished the third and final sampling area for margin sediment — North Bay — with stations located in San Pablo Bay, Carquinez Strait, and Suisun Bay. A report summarizing the results of this study, and comparing them to the results from the margin areas of Central and South Bay, will be completed in 2022.

6 Small Tributaries
Watershed Load Modeling, Phases 2 and 3

In Phase 1 of development of a Watershed Dynamic Model (WDM), the hydrology module was calibrated and completed in 2020 and serves as a solid foundation for the sediment (2021) and pollutant model development (2022 and beyond), beginning with PCBs and mercury. A report on the sediment module will be available in 2022.

7 Nutrients
Trends Analysis

A method for assessing trends and the uncertainty around those trends was recently developed using the USGS 30-year time series for chlorophyll in the Bay. The next steps for this project are to apply that method to gross primary productivity and dissolved oxygen datasets to evaluate differences in water quality trends in the Bay through space and time, and to identify the factors that may be driving those changes. A report will be available in early 2022.

8 PCBs
Passive Sampler Assessment of the Spatial Distribution of PCBs

This study is assessing the loading and spatial distribution of PCBs in the Steinberger Slough/Redwood Creek Priority Margin Unit to address information gaps in the conceptual model for this area. Passive sampling device (PSD) measurements complemented sediment measurements to evaluate the spatial pattern of PCB concentrations in the surface and subsurface sediment. A report will be available in early 2022.

9 Sediment
Bay Sediment Conceptual Model

This project will produce a detailed conceptual model that incorporates recent monitoring and modeling results and working hypotheses of sediment dynamics within the Bay and between subembayments. The conceptual model will inform policy decisions and build frameworks for management, monitoring, and numeric modeling. A report will be available in 2022.

10 The Pulse of the Bay 2022
50th Anniversary of the Clean Water Act

The Pulse makes the most important information available on water quality in the Bay accessible to water quality managers, decision-makers, scientists, and the public. Themes in the 2022 Pulse will include the 50th anniversary of the Clean Water Act and adapting to new challenges such as emerging contaminants, climate change, and COVID. The Pulse will be published at the time of the RMP Annual Meeting in October 2022.

PROGRAM OVERSIGHT

Collaboration and adaptation in the RMP are achieved through the engagement of stakeholders and scientists in frequent committee and workgroup meetings

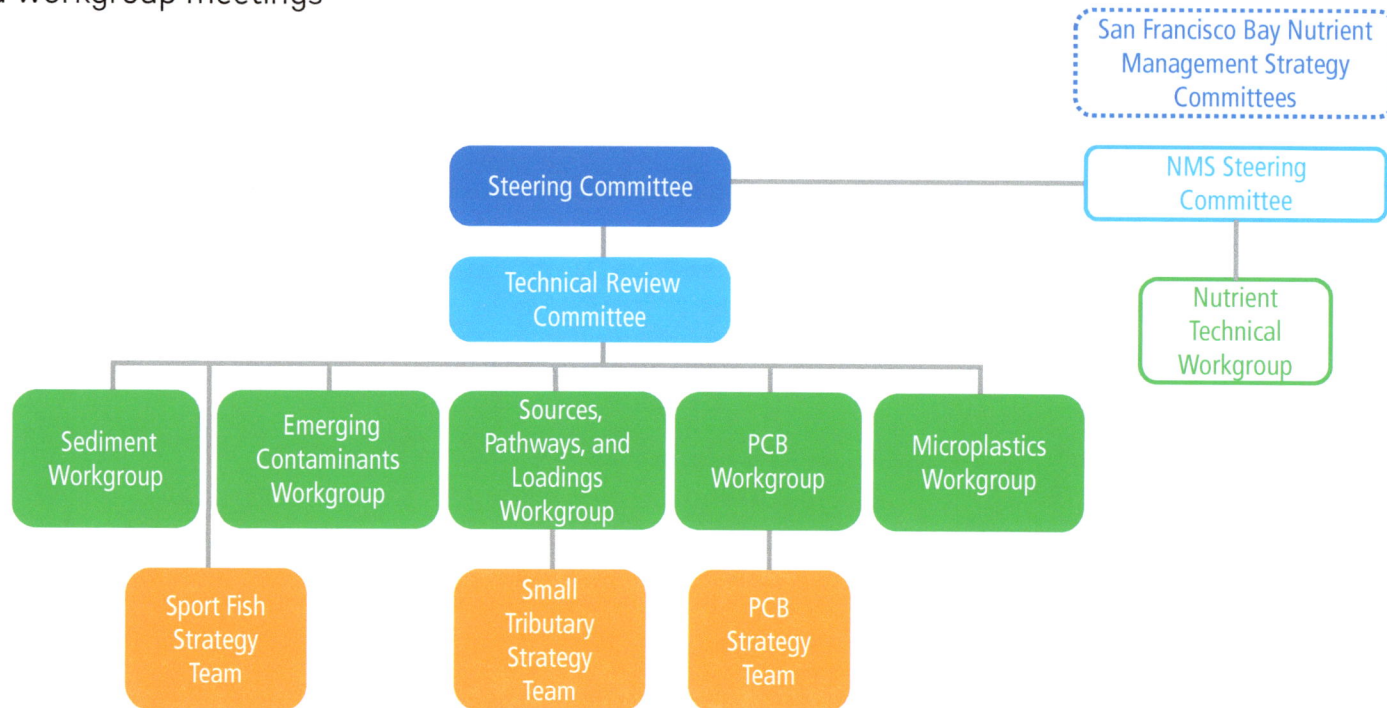

The Steering Committee consists of representatives from discharger groups (wastewater, stormwater, dredging, industrial) and regulatory agencies (Regional Water Board and U.S. Army Corps of Engineers). The Steering Committee determines the overall budget and allocation of program funds, tracks progress, and provides direction to the Program from a manager's perspective.

Oversight of the technical content and quality of the RMP is provided by the **Technical Review Committee** (TRC), which provides recommendations to the Steering Committee.

Workgroups report to the TRC and address the main technical subject areas covered by the RMP. The Nutrient Technical Workgroup was established as part of the committee structure of a separate effort—the Nutrient Management Strategy—and makes recommendations to the RMP committees on the use of the RMP funds that support nutrient studies. The workgroups consist of regional scientists and regulators and invited scientists recognized as authorities in the field. The workgroups directly guide planning and implementation of special studies.

RMP strategy teams constitute one more layer of planning activity. These stakeholder groups meet as needed to develop long-term RMP study plans for addressing high priority topics.

PROGRAM MANAGEMENT

RMP FEES BY SECTOR: 2021

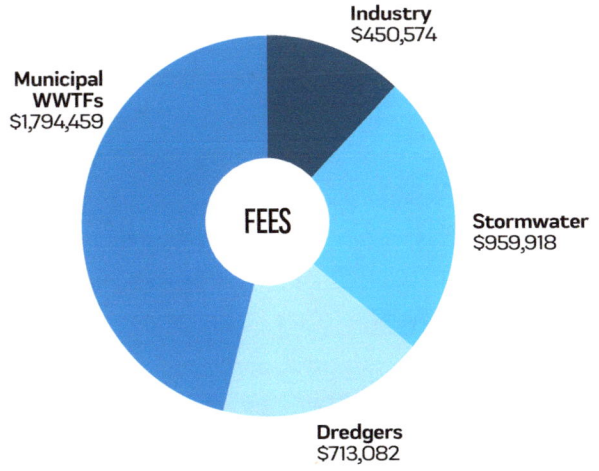

FEES

- Industry $450,574
- Municipal WWTFs $1,794,459
- Stormwater $959,918
- Dredgers $713,082

The fees target for 2021 was $3.92 million.

RMP EXPENSES: 2021

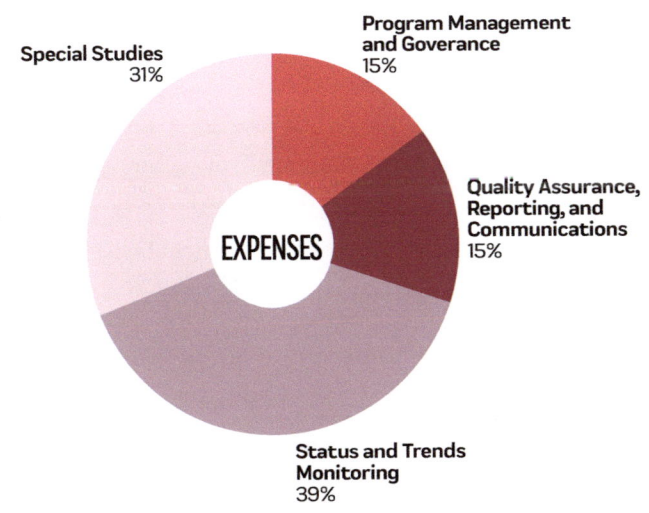

EXPENSES

- Special Studies 31%
- Program Management and Goverance 15%
- Quality Assurance, Reporting, and Communications 15%
- Status and Trends Monitoring 39%

COMMUNICATIONS

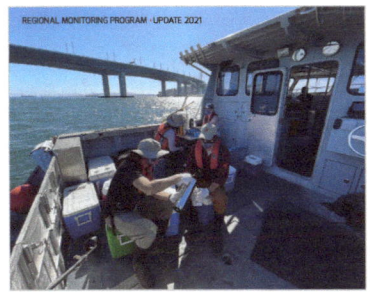

Includes the *RMP Update*, Annual Meeting, Multi-Year Plan, Estuary News articles, the RMP website, technical reports, journal publications, newsletter, oral presentations, posters, and media outreach.

PROGRAM MANAGEMENT AND GOVERNANCE

Includes internal coordination (staff management), committee and workgroup meetings, coordination with Program participants, external coordination with related groups, program planning, contract and financial management, and workgroup and peer review coordination.

DATA MANAGEMENT AND QUALITY ASSURANCE

The RMP database contains approximately 2 million records generated since the Program began in 1993. Web-based data access tools include user-defined queries, data download and printing functionality, maps of sampling locations, and visualization tools.

cd3
CONTAMINANT DATA DISPLAY & DOWNLOAD
cd3.sfei.org

FEATURED PROJECT
RMP MONITORING OF CONTAMINANTS IN SAN FRANCISCO BAY FISH: 2019

BY DR. JAY DAVIS, SFEI

The degree of contamination of San Francisco Bay fish is one of the most important measures of Bay water quality. One of the primary goals of the Clean Water Act is to make all US waters fishable. Fish can be an important part of a healthy, well-balanced diet. They provide an excellent source of protein and vitamins, and are a primary dietary source of heart-healthy omega-3 fatty acids. Fishing on the Bay and consumption of Bay fish are popular and important activities that are enjoyed by many, and many depend heavily on Bay-caught fish in their diets for cultural reasons or subsistence.

◄ Fishing near Vallejo. Photograph by Joe Clark, All Positives Possible
https://www.allpositivesp.org/

Unfortunately, the Bay is not entirely fishable. Contaminants that made their way from land to the Bay decades ago persist and cause concentrations in some Bay fish to be above safe levels. An advisory issued by the California Office of Environmental Health Hazard Assessment, recommending limited consumption of fish from the Bay, has been in place since the 1990s. The advisory recommends no consumption of any surfperch species due to high concentrations of PCBs, and in addition no consumption of striped bass, white sturgeon, brown smoothhound shark, and leopard shark by the sensitive population (women 18-49 years and children 1-17 years) due to high levels of mercury and PCBs.

The San Francisco Bay Regional Water Quality Control Board has established TMDL cleanup plans for mercury and PCBs, two of the contaminants of greatest concern in the Bay, and concentrations in fish are the key benchmarks in these TMDLs.

Progress toward the goal of a fully fishable Bay has been slow. The RMP has been monitoring contaminants in Bay fish since 1997, following up on a pilot study conducted by the Bay Protection and Toxic Cleanup Program in 1994. Over this time span, mercury and PCBs have shown only modest signs of long-term decline. Evidence for declines has been stronger for other contaminants of concern, such as PBDEs and dioxins.

The RMP performs an extensive survey of contaminants in Bay sport fish once every five years. In April 2021, the RMP published a report on the most recent sampling round, which was conducted in 2019. This article presents a

The RMP performs an extensive survey of contaminants in Bay sport fish once every five years

brief overview of the findings for several of the main contaminants of concern in Bay fish: mercury, PCBs, dioxins, PBDEs, and PFAS. Additional details and information are available in the full report.

Mercury, PCBs, dioxins, selenium, polybrominated diphenyl ethers (PBDEs), and per- and polyfluoroalkylated substances (PFAS) were analyzed in 1,306 fish, representing 16 species collected at 13 locations in San Francisco Bay. Fish species were selected based on a number of criteria, including being popular for consumption, sensitive indicators of contaminant accumulation, and wide distribution; representing different exposure pathways (benthic versus pelagic); and having been monitored in the past.

Contaminant concentrations were compared to numeric human health thresholds (advisory tissue levels, or ATLs) established by OEHHA for mercury, PCBs, selenium, and PBDEs. Results were also compared to regulatory thresholds for mercury, PCBs, and selenium, which have been established in TMDL regulations by the San Francisco Bay Regional Water Quality Control Board.

The 2019 survey addressed some of the data gaps identified by OEHHA relating to developing more extensive consumption advice for the Bay. Multiple samples were analyzed for bat rays, northern anchovy, Pacific herring, brown rockfish, and staghorn sculpin. Data gaps remain for diamond turbot, starry flounder, and monkeyface prickleback, where only one sample of each species was analyzed, and other species of interest that were not analyzed (Pacific sardine, cabezon, Pacific sanddab, and petrale sole).

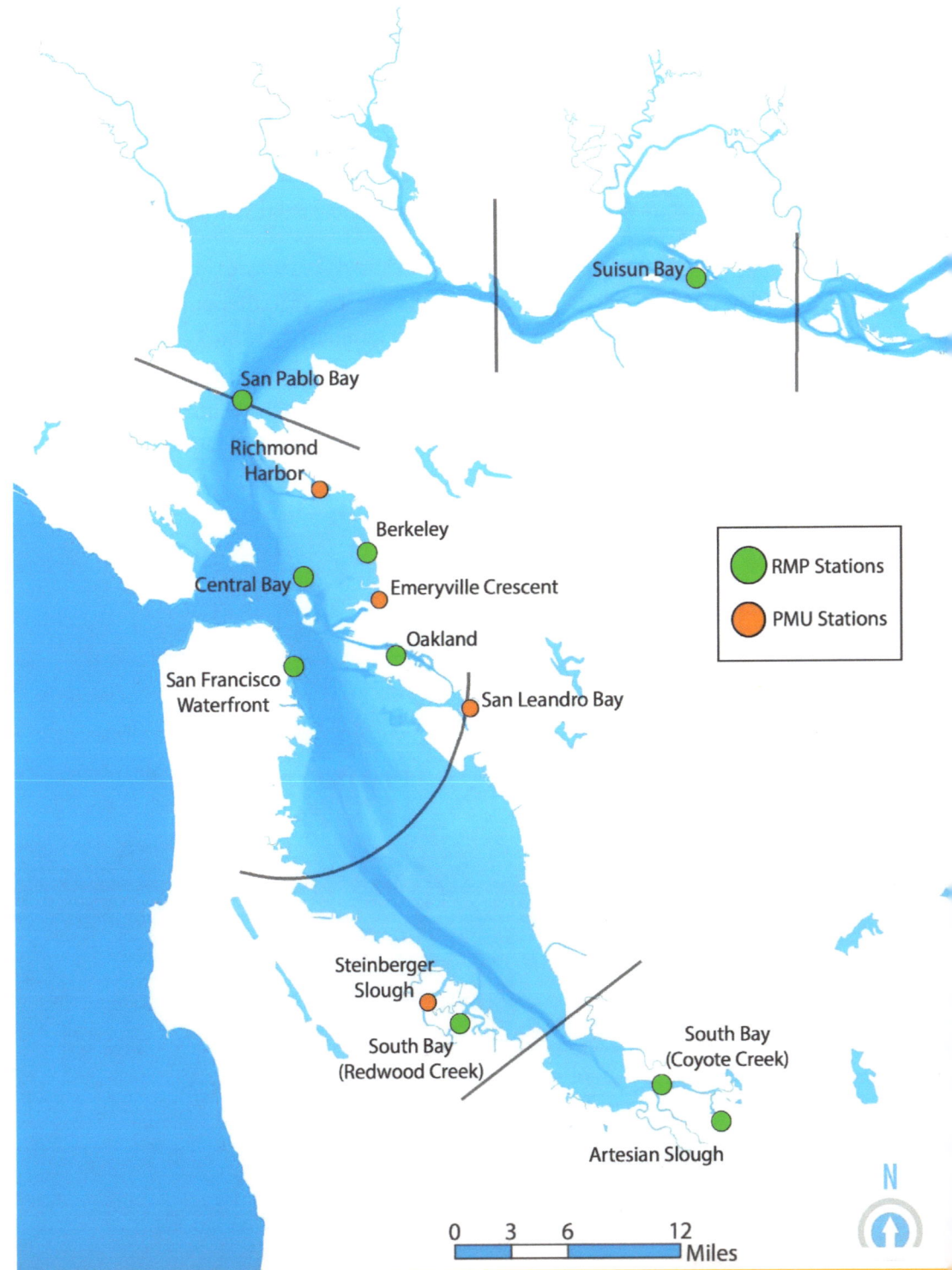

MERCURY

The fish advisory for the Bay is primarily driven by human health risks due to exposure to mercury and PCBs. The 2019 data show that mercury and PCB concentrations remain above thresholds and are widespread, indicating these contaminants continue to pose the greatest human and wildlife health risks.

Mercury concentrations continue to exceed thresholds of concern in Bay sport fish. The average mercury concentrations in bat rays (0.83 ppm) and striped bass (0.46 ppm) exceeded the no consumption ATL (for the sensitive population) of >0.44 ppm, and a few white croaker composites and individual largemouth bass exceeded this threshold as well, as did one individual white sturgeon and one sample of shiner surfperch. Lower concentrations were measured in other popularly consumed sport fish species. White croaker, white sturgeon, and diamond turbot had average concentrations that fell within the one serving/week ATL range (>0.15-0.44 ppm) for the sensitive population; shiner surfperch, California halibut, brown rockfish, starry flounder, and jacksmelt averages were in the two serving/week range (range = >0.07- 0.15 ppm); northern anchovy were in the three serving/week range (>0.055-0.07 ppm); Pacific herring, staghorn sculpin, white surfperch, and monkeyface prickleback averages fell below the three serving/week range (<0.055 ppm).

> *Mercury concentrations continue to exceed thresholds of concern in Bay sport fish*

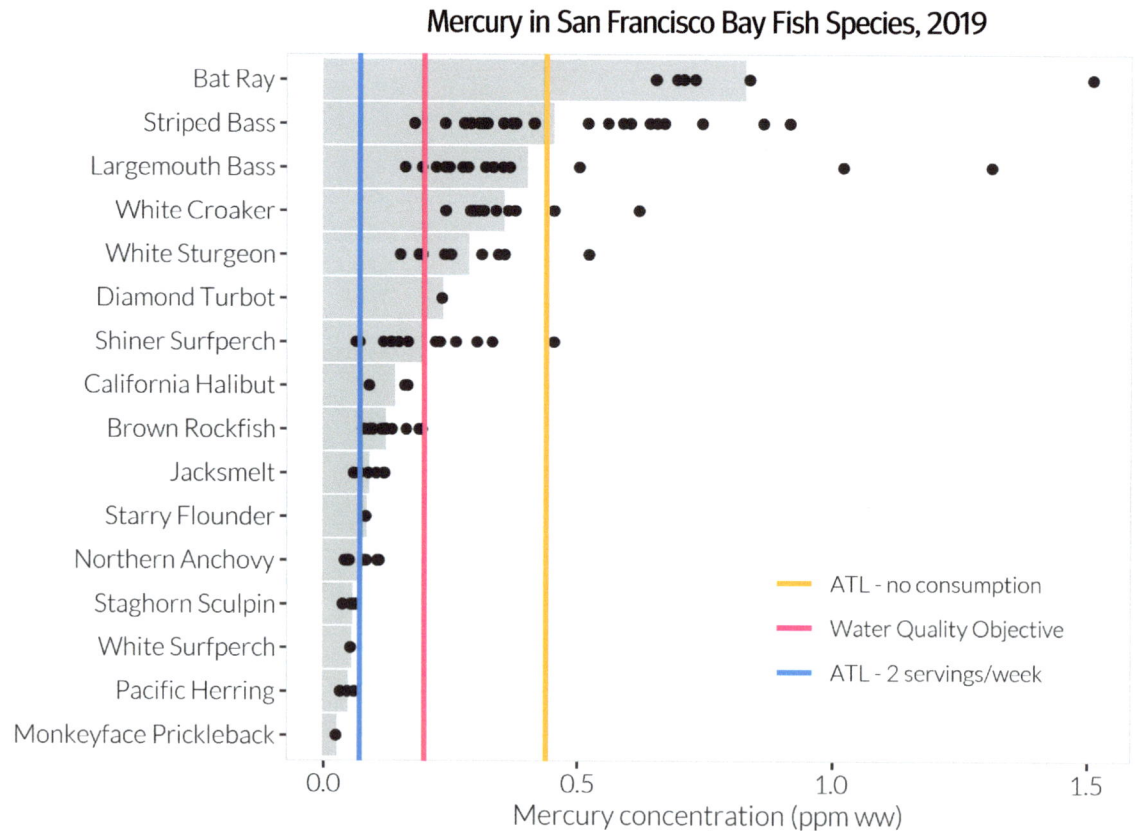

Mercury in San Francisco Bay Fish Species, 2019

Legend:
- ATL - no consumption
- Water Quality Objective
- ATL - 2 servings/week

Mercury concentration (ppm ww)

Mercury concentrations in Bay fish are not showing signs of long-term decline. Striped bass is the most important indicator species for mercury in the Bay, due to its popularity for consumption and the high concentrations of mercury that it accumulates. Striped bass from the Bay have the highest average mercury concentration measured for this species in US estuaries.

A relatively extensive historical dataset exists for Bay striped bass, allowing for the evaluation of trends over 44 years, from 1971 to 2019. In 2019, the average mercury concentration was not significantly different from the average in 1971. Furthermore, the overall long-term trend line does not indicate a change over the 44-year period.

Mercury in Striped Bass, 1971-2019

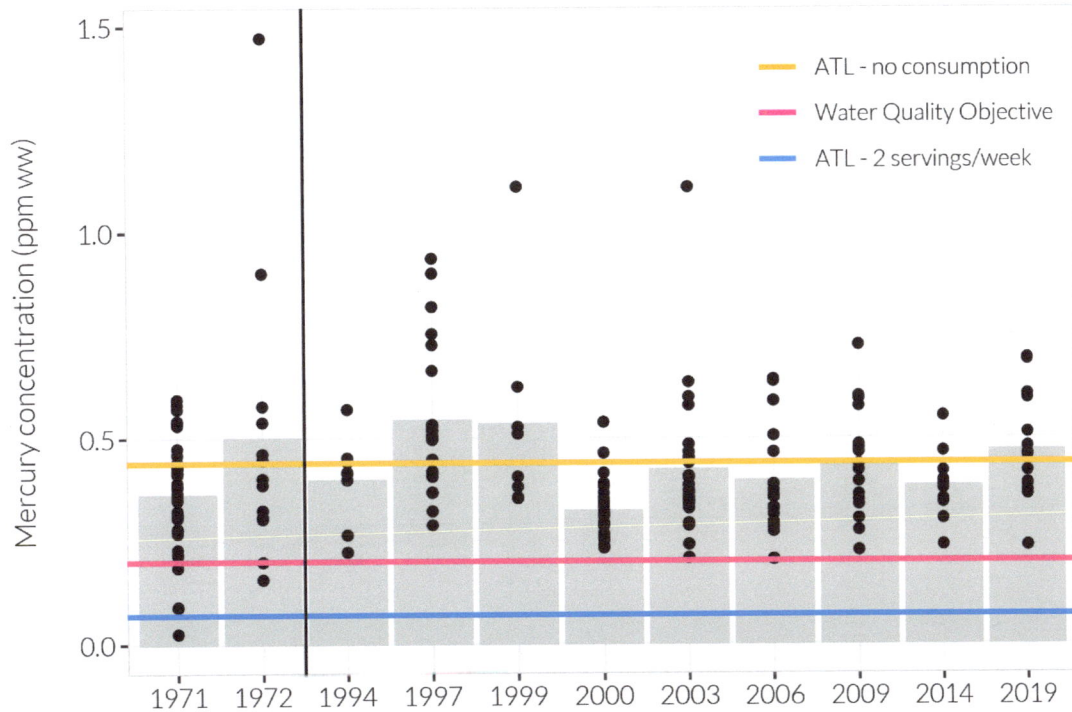

Legend:
- ATL - no consumption
- Water Quality Objective
- ATL - 2 servings/week

Y-axis: Mercury concentration (ppm ww)
X-axis years: 1971, 1972, 1994, 1997, 1999, 2000, 2003, 2006, 2009, 2014, 2019

Fisherman at Municipal Pier in Aquatic Park in San Francisco. ▼ Photograph by Nicole David.

FOOTNOTE: Bars indicate average concentrations. Points represent individual fish, with the exception of six composite samples (3 fish each) analyzed in 2014. All plotted points are 60 cm length-adjusted. The 2014 data do not include fish collected in Artesian Slough, and the 2019 data do not include fish collected in South Bay (Coyote Creek); these areas reflect unique mercury sources and were collected only in those years. Data were obtained from CDFW historical records (1971-1972), the Bay Protection and Toxic Cleanup Program (1994), a CalFed-funded collaborative study (1999 and 2000), and the Regional Monitoring Program (1997, 2000, 2003, 2006, 2009, 2014, and 2019). The colored lines indicating ATL thresholds show the lower end of ATL ranges for the sensitive population.

PCBs

PCBs, along with mercury, are a main driver of the consumption advisory for the Bay.

PCB concentrations in Bay sport fish remain high and continue to exceed thresholds of concern, including both human consumption thresholds and water quality regulatory thresholds. The highest species average PCB concentration was for shiner surfperch (220 ppb), exceeding all thresholds, with concentrations of some composites over two times greater than the no consumption ATL (>120 ppb), and a maximum concentration of 400 ppb.

Northern anchovy, an indicator species for wildlife exposure, had the second highest average concentration (110 ppb).

More moderate concentrations were measured in other species, ranging from 81 ppb in largemouth bass to 0.5 ppb in monkeyface prickleback. Ten of the 16 species measured had average concentrations in exceedance of the numeric target from the TMDL (10 ppb). The other six species below the numeric target included one commonly-consumed species (California halibut), and five other less commonly consumed species (brown rockfish, diamond turbot, monkeyface prickleback, Pacific herring, starry flounder).

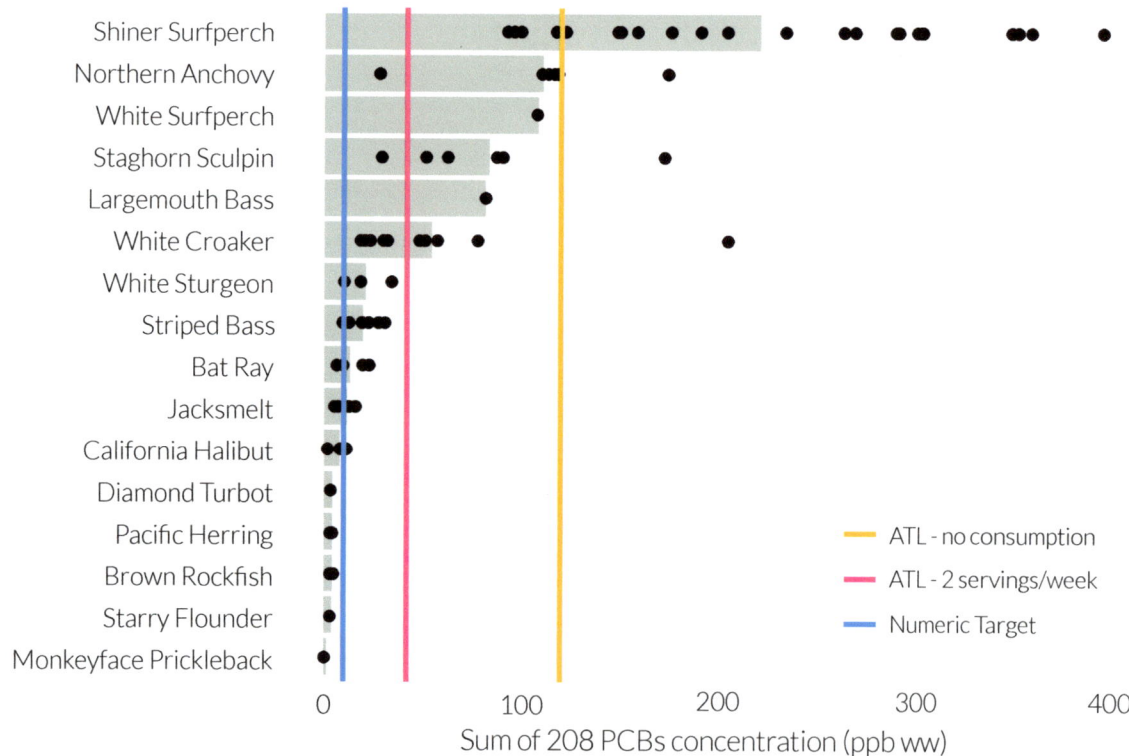

PCBs in San Francisco Bay Fish Species, 2019

> *PCB concentrations in Bay sport fish remain high and continue to exceed thresholds of concern*

FOOTNOTE: Bars indicate average concentrations. Points represent individual samples (either composites or individual fish). The colored lines indicating ATL thresholds show the lower end of the advisory tissue level ranges.

PCB concentrations vary significantly across the four long-term monitoring stations (Berkeley, Oakland Harbor, San Francisco Waterfront, and South Bay). Oakland Harbor remains the region of highest concern, although San Francisco Waterfront and South Bay also had average concentrations above the no consumption ATL of 120 ppb in this round of sampling.

Shiner surfperch are excellent indicators of spatial variability in PCB concentrations in the Bay. The spatial distribution of PCB contamination observed in shiner surfperch at the long-term stations was consistent with patterns observed in earlier rounds of sampling. One difference from prior rounds was that shiner surfperch were not collected from the San Pablo Bay station, which historically has consistently had the lowest average PCB concentrations. As in prior rounds, PCB concentrations were higher in Oakland Harbor (280 ppb) than at the other S&T stations, with a statistically significant difference between Oakland Harbor and the lowest average concentration at Berkeley (94 ppb). Average concentrations were intermediate at South Bay (Redwood Creek) (180

ppb) and at the San Francisco Waterfront (180 ppb) and were not significantly different from any of the other stations.

Two additional areas sampled as part of the PMU special study (Richmond Harbor and San Leandro Bay) had relatively high average concentrations that, like Oakland Harbor, were significantly different from Berkeley (the station with the lowest average concentration). The average concentration at San Leandro Bay (350 ppb) was even higher than the average at Oakland Harbor, while the average at Richmond Harbor (230 ppb) was the third highest overall behind San Leandro Bay and Oakland Harbor.

Significant spatial variation was observed within Richmond Harbor between stations that were only 0.9 km apart. Mean concentrations for the Santa Fe Channel (269 ppb) and Lauritzen Channel (306 ppb) were significantly higher than the mean for the Main Channel (113 ppb). This dataset provides a clearer understanding of the high site fidelity of shiner surfperch.

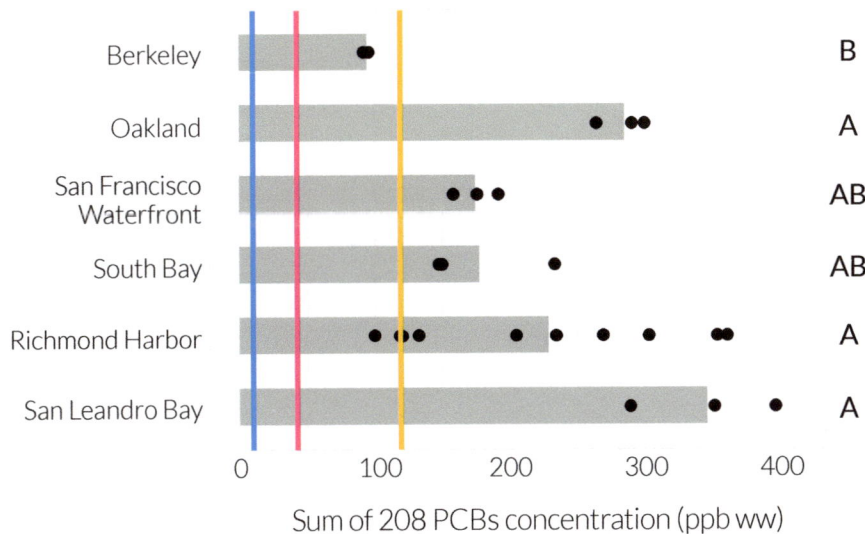

PCBs in Shiner Surfperch by Location, 2019

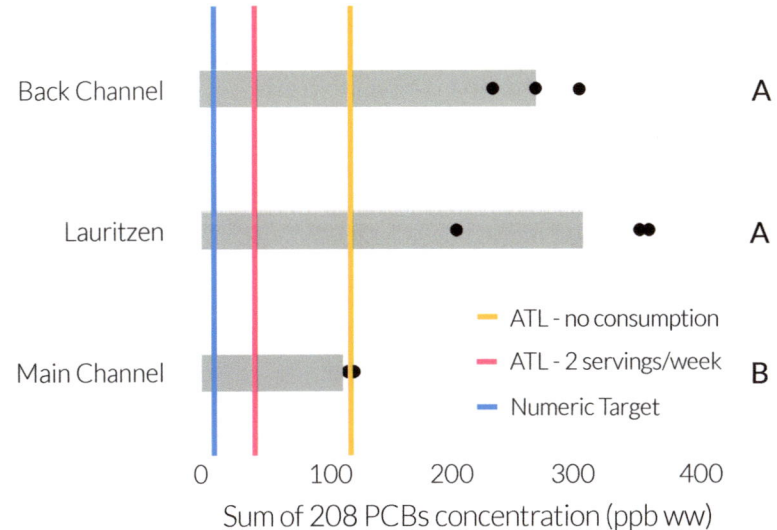

PCBs in Shiner Surfperch in Richmond Harbor, 2019

FOOTNOTE (both graphs): Bars indicate average concentrations. Points represent composite samples. Colored lines indicating ATL thresholds show the lower end of the ATL ranges. Locations labeled with the same letter did not have significantly different means.

PCBs

Although PCB concentrations in shiner surfperch (the primary indicator species) were generally higher in 2019 than in the prior round of sampling, there are some possible signs of long-term decline. Overall, the rate of PCB decline in the Bay is slow at best, and continued monitoring is needed for a more definitive assessment.

Although the long-term wet weight time series at individual stations through 2014 suggested possible declining trends, higher concentrations were observed across the stations in 2019 that weakened these patterns. At each of the long-term stations sampled in 2019, concentrations were higher than in 2014, and the differences were substantial for Berkeley, San Francisco Waterfront, and South Bay. For San Francisco Waterfront and South Bay, the concentrations went from being well below the 120 ppb no consumption ATL in 2014 to above this threshold in 2019. While the larger number of congeners analyzed in 2019 contributed to the higher values, the differences at Berkeley, San Francisco Waterfront, and South Bay were larger than the approximate 15% increase the added congeners would cause.

Another driver of the higher concentrations in 2019 was relatively high lipid (fat) content of the tissue in 2019. PCBs accumulate primarily in lipid. Taking lipid content of the samples into account (data not shown), there appear to be indications of long-term declines, most clearly at the Berkeley station. Overall, the wet weight and lipid weight PCB data for shiner surfperch suggest that ambient PCB concentrations in the Bay have not declined substantially Bay-wide between 1994 and 2019.

PCBs in Shiner Surfperch by Location, 1994-2019

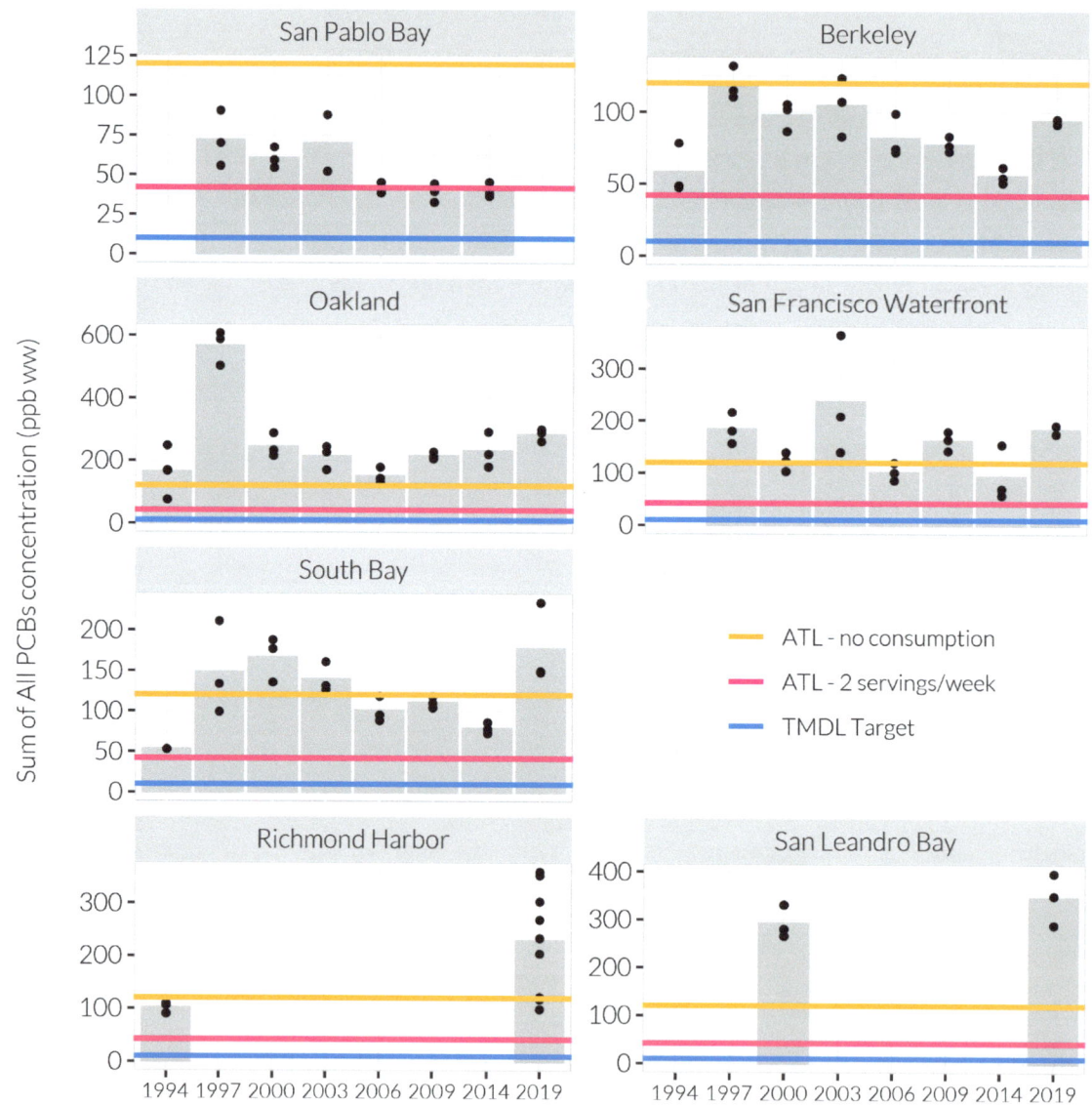

FOOTNOTE: Bars indicate average concentrations. Points represent composite samples with 20 fish in each composite. Data shown are the sum of PCBs for all congeners analyzed; the number analyzed varied from 47 in 1994 to 52 in 2014, and then increased to 209 in 2019. The colored lines indicating ATL thresholds show the lower end of the advisory tissue level ranges.

The PCB concentrations observed in white croaker in 2019, on the other hand, were the lowest yet observed, suggestive of a possible long-term decline. RMP assessment of long-term trends in PCBs has historically relied on both shiner surfperch and white croaker data. While shiner surfperch, due to their high site fidelity, represent exposure in specific locations, white croaker range more widely and provide a more spatially integrated index of regional contaminant exposure in the Bay food web. The Bay-wide average sum of 40 PCBs concentration for white croaker on a wet weight basis in 2019 was 45 ppb, less than half of the concentration measured in 2009, and far below the average concentrations measured for skin-on fillets in the rounds before 2009.

As for shiner surfperch, much of the variation across the years is due to variation in the lipid content of the tissues. The lipid variation in white croaker over the years was further heightened by a switch from analysis of fillets with skin (which has high lipid content) from 1997-2006 to fillets without skin from 2009 to the present. However, even on a lipid-normalized basis the 2019 average concentration was also distinctly lower than those observed in previous years. Continued monitoring is needed, however, to establish whether the white croaker data are indeed signaling a trend rather than merely high year-to-year variation.

PCBs in White Croaker, 1997-2019

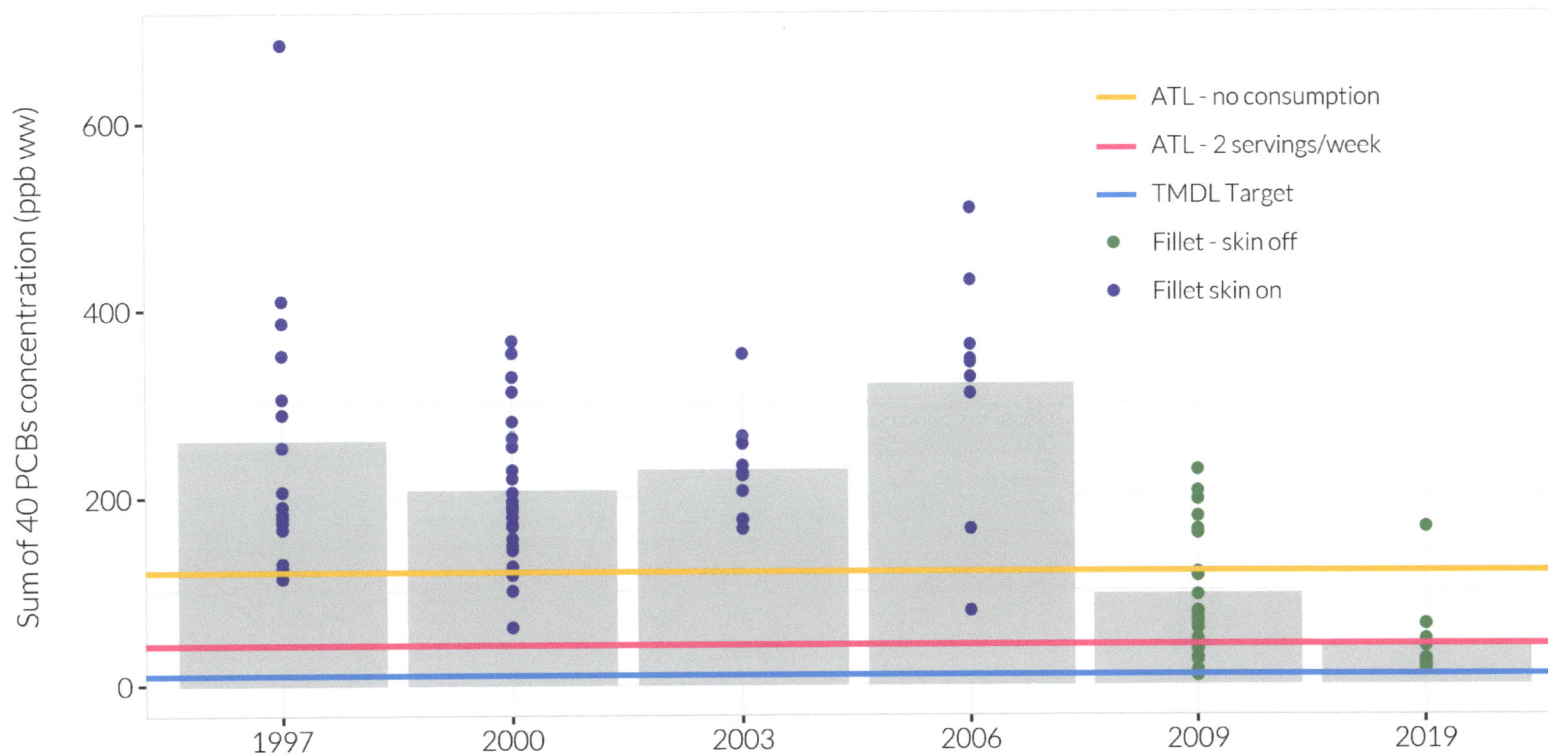

The PCB concentrations observed in white croaker in 2019 were the lowest yet observed, suggestive of a possible long-term decline

FOOTNOTE: Bars indicate average concentrations. Points represent composite samples with five fish in each composite.

DIOXINS

Dioxins are at concentrations of potential concern in the Bay, but neither a Water Board regulatory target nor OEHHA advisory tissue levels have been established. As part of the PCB TMDL, because some PCBs have the same mechanism of toxicity as dioxins, the Water Board calculated a fish tissue screening level for dioxins of 0.14 pptr (parts per trillion) for the assessment of risk to human health.

Dioxin concentrations in Bay fish remain above the Water Board screening level, and are still particularly high in Oakland Harbor. However, there are signs of possible decline in both of the key indicator species: shiner surfperch and white croaker.

In shiner surfperch, concentrations appear to be progressively decreasing across all of the monitoring stations except Oakland Harbor, although the decline is not statistically significant at any of the monitoring stations.

Dioxins in Shiner Surfperch By Location, 1994-2019

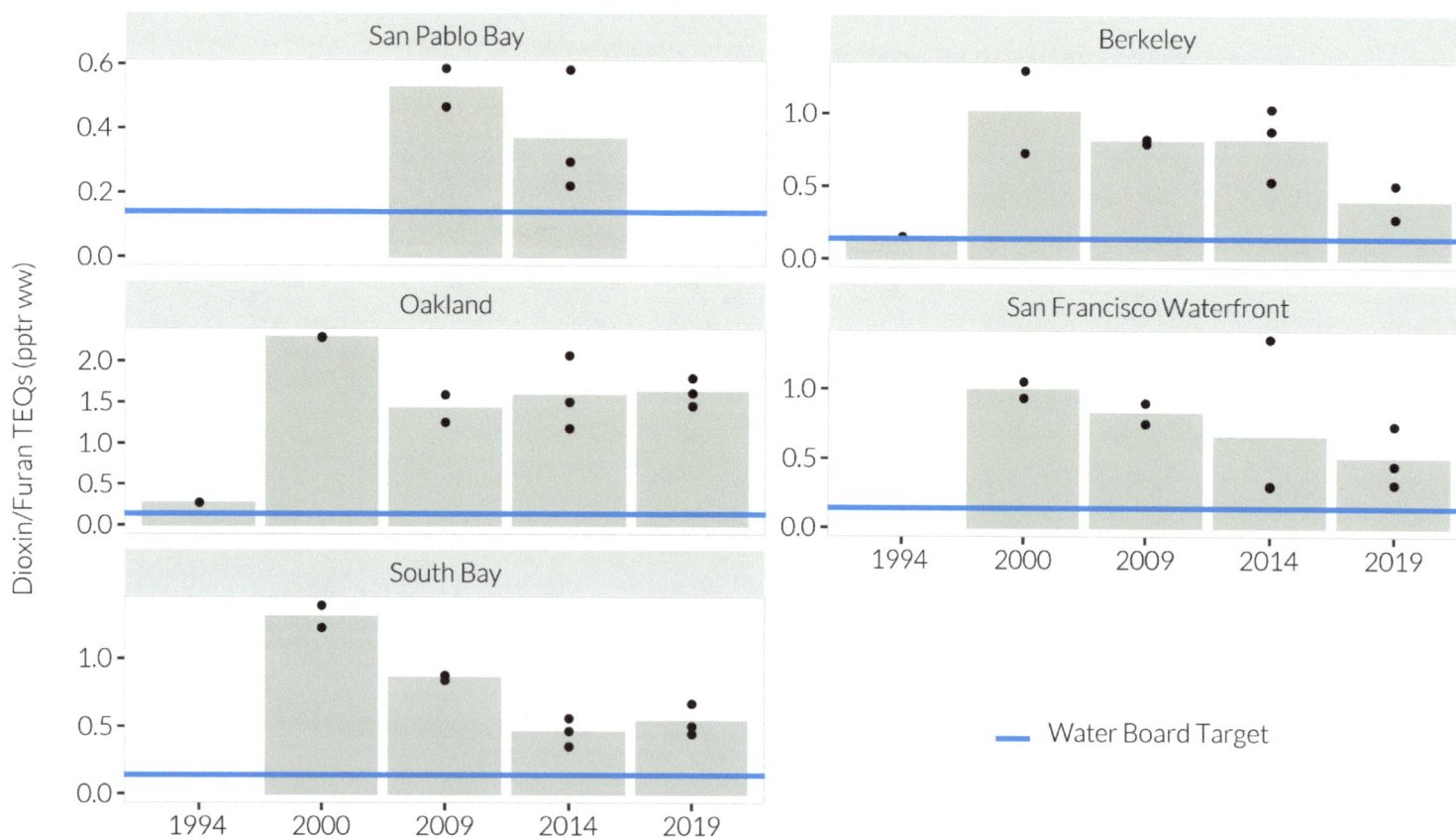

FOOTNOTE: Bars indicate average concentrations. Points represent composite samples with 20 fish in each composite.

In white croaker, the concentrations in 2019 were sharply lower than the last year of comparable data in 2009, and only slightly above the screening level.

Continued monitoring of shiner surfperch and white croaker is needed to establish whether these possible trends reach a point of statistical significance and are signs of actual long-term declines.

In white croaker, dioxin concentrations in 2019 were sharply lower than the last year of comparable data in 2009, and only slightly above the screening level

Dioxins in White Croaker, 1994-2019

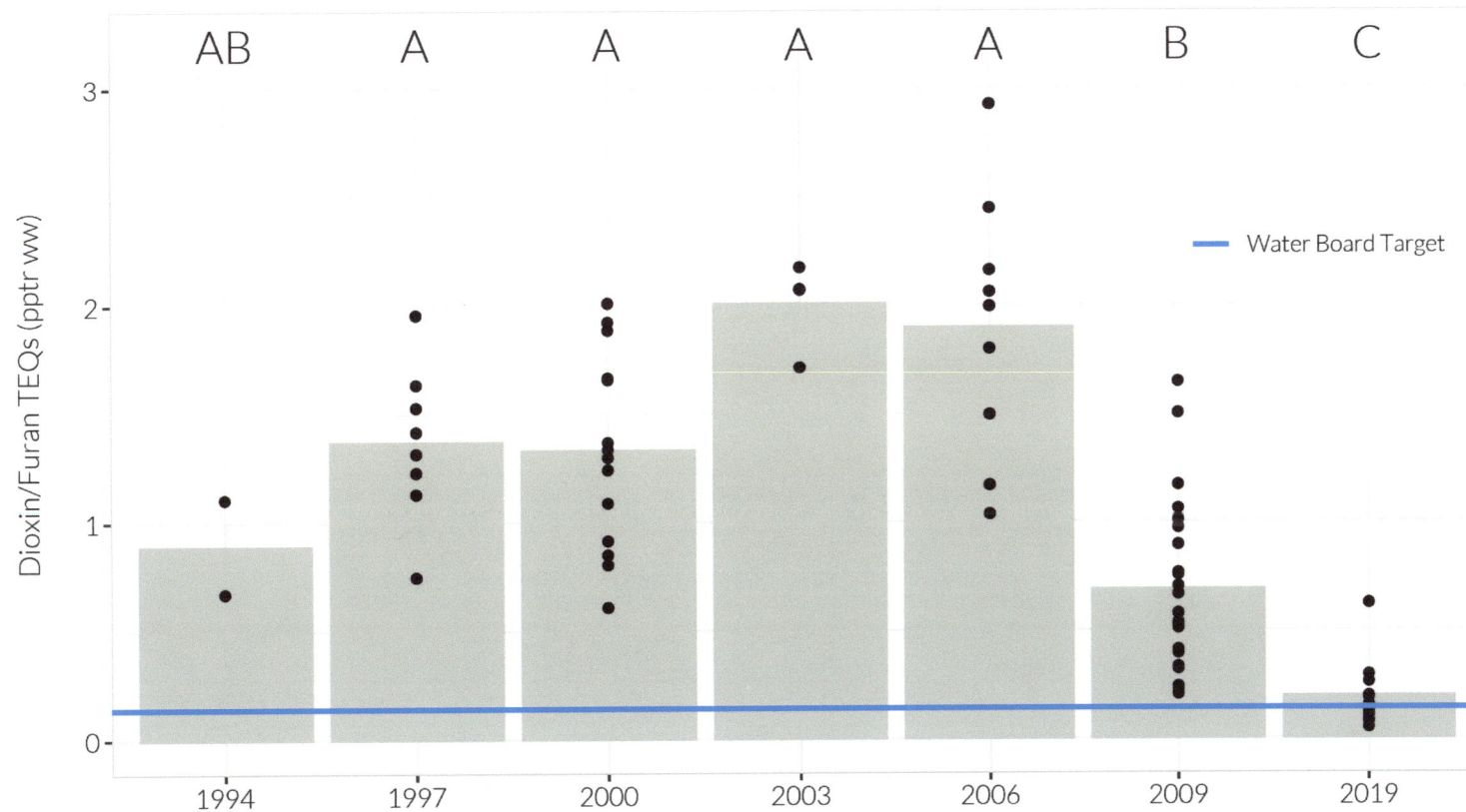

FOOTNOTE: Dioxin TEQsPCDD/PCDF (pptr ww) in white croaker in San Francisco Bay, 1994-2019. Bars indicate average concentrations. Points represent composite samples with 5 fish in each composite. The Water Board screening level (0.14 pptr) is non-regulatory. Years labeled with the same letter did not have significantly different means.

PBDEs

The 2019 PBDE data provide further evidence of the decline of PBDEs in Bay sport fish following the PBDE bans and phase-outs in the mid-2000s. The rate of decline has levelled off in recent years, but the current concentrations are well below the lowest OEHHA advisory tissue level for the protection of human health (45 ppb).

Statistically significant declines were observed for the Bay as a whole and consistently across nearly all of the individual monitoring stations; the decline was not significant at Oakland (data not shown).

The RMP Emerging Contaminant Workgroup's monitoring plan calls for one more round of PBDE measurement in 2024 for further confirmation of the long-term decline.

PBDEs in Shiner Surfperch, 2003-2019

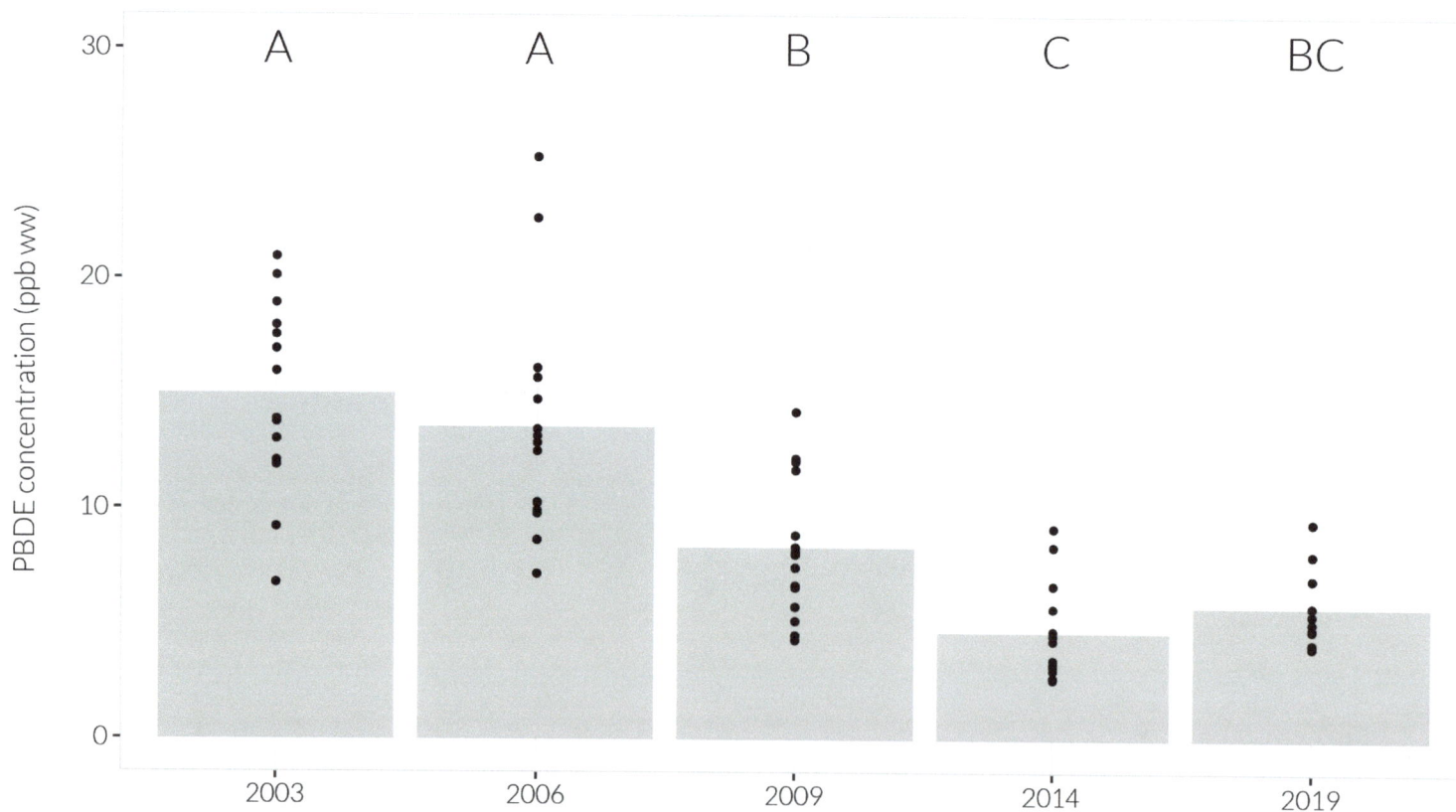

FOOTNOTE: Bars indicate average concentrations. Points represent composite samples with 20 fish in each composite. Years labeled with the same letter did not have significantly different means.

PFAS

No regulatory or human health thresholds have been established for PFAS in San Francisco Bay fish. Concentrations in Bay fish, however, particularly in the South Bay region, are persisting over time at levels that exceed consumption advisory thresholds that have been established by other states. The monitoring conducted to date for PFAS in fish has been inconsistent and limited in scope, hindering evaluation of spatial patterns and long-term trends. More intensive monitoring is warranted to track long-term trends, understand spatial variation across Bay regions, and more firmly characterize concentrations for comparison to thresholds. The Lower South Bay appears to be a region of particular concern, especially Artesian Slough, which is in close proximity to the outfall for the San Jose-Santa Clara Regional Wastewater Facility.

PFAS Concentrations in Regions of San Francisco Bay, 2009-2019

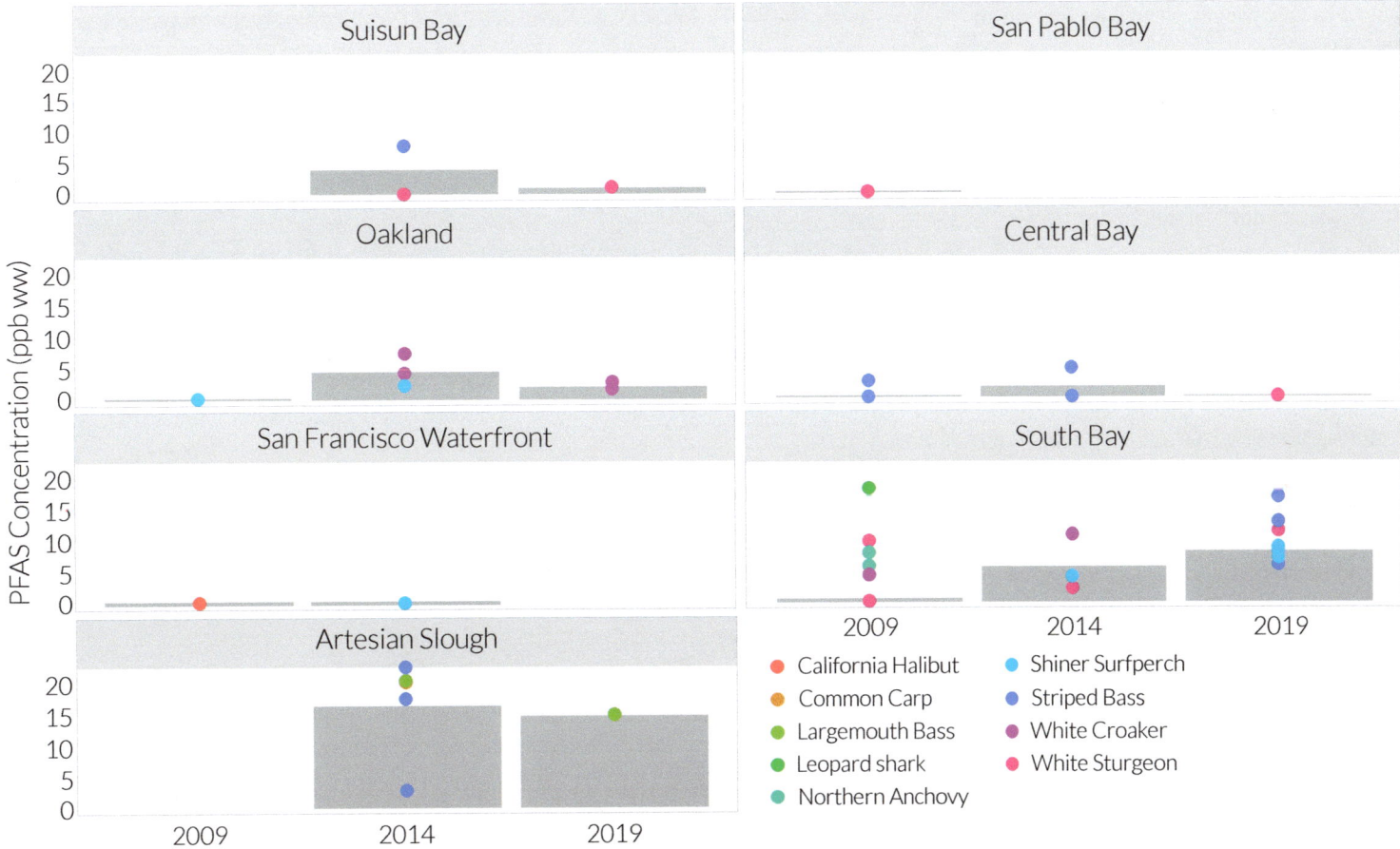

FOOTNOTE: Bars indicate average concentrations. Points represent composite samples of indicated fish species. The number of PFAS analytes and MDLs varied across years.

WHAT'S NEXT?

Tribal and Subsistence Beneficial Use Designations: Interest has increased recently in the impacts of fish contamination on groups with high consumption rates, including subsistence fishers and tribes. In 2017 the State Water Board established three new beneficial use definitions for use by the State and Regional Water Boards in designating Tribal Traditional Culture, Tribal Subsistence Fishing, and Subsistence Fishing beneficial uses to inland surface waters, enclosed bays, or estuaries. The nine regional water boards are at various stages of considering whether to designate these uses for water bodies in their regions.

Survey of Subsistence Fishers: Related to this process, the San Francisco Bay Regional Water Board plans to fund a pilot consumption study to obtain updated and expanded information on fish consumption to assess whether existing objectives and thresholds for mercury and PCBs are protective of subsistence fishers. The pilot study will characterize consumption of Bay-caught fish by subsistence fishers in the Carquinez Strait region.

Community-Guided Statewide Monitoring: In another related effort, the State Water Board's Surface Water Ambient Monitoring Program has launched a statewide initiative to engage with communities with high consumption rates to help direct monitoring effort toward the locations and species that those communities target most. The initiative is starting in the San Diego region, then successively working through the other eight regions.

Enhanced Assessment of PFAS: Another development on the horizon is an enhanced assessment of PFAS in Bay fish. The next round of RMP monitoring in 2024 will include more thorough monitoring of PFAS to provide better information on spatial patterns and long-term trends. In addition, OEHHA is currently evaluating the toxicity of PFAS and may be able to develop advisory tissue levels for one or more PFAS chemicals as they complete the evaluations. A first step may involve interim consumption advice for hotspots like Artesian Slough.

Continued Tracking of Contaminants in Bay Fish: The next round of monitoring in 2024 will also include continued monitoring of mercury, PCBs, and other contaminants to track trends, and will again attempt to target additional species to support enhanced consumption advice for the Bay. In cooperation with SWAMP, this may include additional monitoring to characterize the locations and species that subsistence fishers depend upon most.

> *The next round of RMP monitoring in 2024 will include more thorough monitoring of PFAS to provide better information on spatial patterns and long-term trends*

Fishing on the eastern shoreline of Central Bay.
Photograph by Shira Bezalel.

RECENT PUBLICATIONS

RMP UPDATE 2020

RMP Update 2020. Davis, J.; Foley, M.; Askevold, R.; Buzby, N.; Chelsky, A.; Dusterhoff, S.; Gilbreath, A.; Lin, D.; Miller, E.; Senn, D.; et al. 2020. SFEI Contribution No. 1008.
https://www.sfei.org/documents/rmp-update-2020

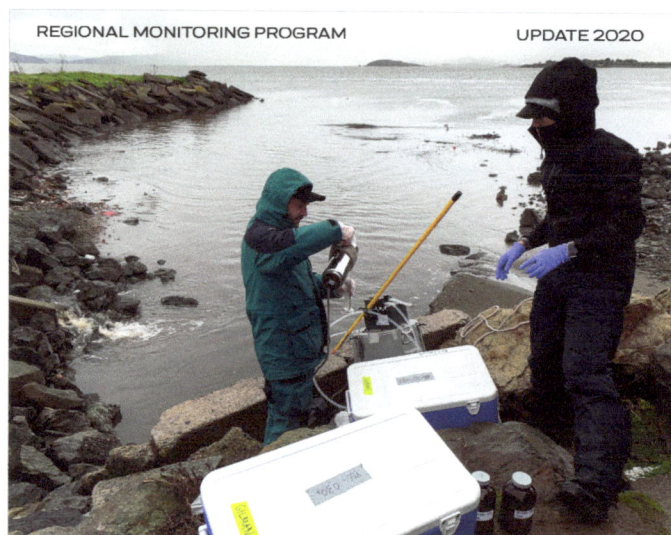

REGIONAL MONITORING PROGRAM   UPDATE 2020

JOURNAL PUBLICATIONS

Think Global, Act Local: Local Knowledge Is Critical to Inform Positive Change When It Comes to Microplastics. Rochman, C. M.; Munno, K.; Box, C.; Cummins, A.; Zhu, X.; Sutton, R. 2020. Environmental Science & Technology. SFEI Contribution No. 1024. https://www.sfei.org/documents/think-global-act-local-local-knowledge-critical-inform-positive-change-when-it-comes

Recommended Best Practices for Collecting, Analyzing, and Reporting Microplastics in Environmental Media: Lessons Learned from Comprehensive Monitoring of San Francisco Bay. Miller, E.; Sedlak, M.; Lin, D.; Box, C.; Holleman, C.; Rochman, C. M.; Sutton, R. 2020. Journal of Hazardous Materials . SFEI Contribution No. 1023. https://www.sfei.org/documents/recommended-best-practices-collecting-analyzing-and-reporting-microplastics-environmental

Methods Matter: Methods for Sampling Microplastic and Other Anthropogenic Particles and Their Implications for Monitoring and Ecological Risk Assessment. Hung, C.; Klasios, N.; Zhu, X.; Sedlak, M.; Sutton, R. 2020. Integrated Environmental Assessment and Management 16 (6). SFEI Contribution No. 1014. https://www.sfei.org/documents/methods-matter-methods-sampling-microplastic-and-other-anthropogenic-particles-and-their

A ubiquitous tire rubber-derived chemical induces acute mortality in coho salmon. Tian, Z.; Zhao, H.; Peter, K.T.; Gonzalez, M.; Wetzel, J.; Wu, C.; Hu, X.; Prat, J.; et al. 2020. Science 371 (6525), 185-189. https://www.science.org/doi/10.1126/science.abd6951

Assessment of emerging polar organic pollutants linked to contaminant pathways within an urban estuary using non-targeted analysis. Overdahl, K.E.; Sutton, R.; Sun, J.; DeStafano, N.J.; Getzinger, G.J., Ferguson, P.L. 2021. Environmental Science: Processes & Impacts. https://doi.org/10.1039/D0EM00463D

Suspended-sediment Flux in the San Francisco Estuary; Part II: the Impact of the 2013–2016 California Drought and Controls on Sediment Flux. Livsey, D.N.; Downing-Kunz, M.A.; Schoellhamer, D.H.; Manning, A. 2020. Estuaries and Coasts. https://doi.org/10.1007/s12237-020-00840-y

Tidal Asymmetry in Ocean-Boundary Flux and In-Estuary Trapping of Suspended Sediment Following Watershed Storms: San Francisco Estuary, California, USA. Downing-Kunz, M.A.; Work, P.A.; Schoellhamer, D.H. 2021. Estuaries and Coasts, pp.1-18.

Urban Stormwater Runoff: A Major Pathway for Anthropogenic Particles, Black Rubbery Fragments, and Other Types of Microplastics to Urban Receiving Waters. Werbowski, L.M.; Gilbreath, A.N.; Munno, K.; Zhu, X.; Grbic, J.; Wu, T.; Sutton, R.; Sedlak, M.D.; Deshpande, A.D.; Rochman, C.M. 2021. ACS ES&T Water. https://doi.org/10.1021/acsestwater.1c00017

Framework for nontargeted investigation of contaminants released by wildfires into stormwater runoff: Case study in the northern San Francisco Bay area. Chang, D.; Richardot, W.; Miller, E.; Dodder, N.; Sedlak, M.; Hoh, E.; Sutton, R. 2021. Integrated Environmental Assessment and Management. SFEI Contribution No. 1044. https://setac.onlinelibrary.wiley.com/doi/10.1002/ieam.4461

PROGRAM PLANNING DOCUMENTS

2021 Bay RMP Multi-Year Plan. Foley, M. 2021. SFEI Contribution No. 1027. San Francisco Estuary Institute: Richmond, CA. https://www.sfei.org/documents/2021-rmp-multi-year-plan

2020 Bay RMP Detailed Workplan and Budget. 2020. SFEI Contribution No. 980. San Francisco Estuary Institute: Richmond, CA. https://www.sfei.org/documents/2020-bay-rmp-detailed-workplan-and-budget

2019 Update to Copper Rolling Average. Yin, J. 2021. 2019 Update to Copper Rolling Average. https://www.sfei.org/documents/2017-update-copper-rolling-average

2019 Update to Cyanide Rolling Average. Yin, J. 2021. 2019 Update to Cyanide Rolling Average. https://www.sfei.org/documents/2017-update-cyanide-rolling-average-0

PRESENTATIONS

Find links to all presentations from the 2020 RMP Annual Meeting on the 2020 Annual Meeting webpage. https://www.sfei.org/events/2020-rmp-annual-meeting

Find links to all presentations from the 2021 RMP Annual Meeting on the 2021 Annual Meeting webpage. https://www.sfei.org/events/2021-rmp-annual-meeting

SUMMARY FOR MANAGERS

Summary for Managers: Non-targeted Analysis of Stormwater Runoff following the 2017 Northern San Francisco Bay Area Wildfires. Miller, E.; Sedlak, M.; Sutton, R.; Chang, D.; Dodder, N.; Hoh, E. 2021. SFEI Contribution No. 1045. San Francisco Estuary Institute: Richmond, CA. https://www.sfei.org/documents/summary-managers-non-targeted-analysis-stormwater-runoff-following-2017-northern-san

TECHNICAL REPORTS

Special Study on Bulk Density. McKnight, K.; Lowe, J.; Plane, E. 2020. SFEI Contribution No. 975. San Francisco Estuary Institute: Richmond, CA. https://www.sfei.org/documents/special-study-bulk-density

Summary of Water Year 2017 precipitation, discharge, and sediment conditions at selected locations in Arroyo de la Laguna watershed, with a focus on Arroyo Mocho. Pearce, S.; McKee, L. 2020. SFEI Contribution No. 912. San Francisco Estuary Institute: Richmond, CA. https://www.sfei.org/documents/summary-water-year-2017-precipitation-discharge-and-sediment-conditions-selected-locations

Conceptual Model to Support PCB Management and Monitoring in the Steinberger Slough/Redwood Creek Priority Margin Unit. Yee, D.; Gilbreath, A.; McKee, L.; Davis, J. 2020. SFEI Contribution No. 1009. San Francisco Estuary Institute: Richmond, CA. https://www.sfei.org/documents/conceptual-model-support-pcb-management-and-monitoring-steinberger-sloughredwood-creek

Pollutants of Concern Reconnaissance Monitoring Progress Report, Water Years 2015 - 2019. Gilbreath, A.; Hunt, J.; Mckee, L. 2020. SFEI Contribution No. 987. San Francisco Estuary Institute: Richmond, CA. https://www.sfei.org/documents/pollutants-concern-reconnaissance-monitoring-progress-report-water-years-2015-2019

Sediment Monitoring and Modeling Strategy. Mckee, L.; Lowe, J.; Dusterhoff, S.; Foley, M.; Shaw, S. 2020. SFEI Contribution No. 1016. San Francisco Estuary Institute: Richmond, CA. https://www.sfei.org/documents/sediment-monitoring-and-modeling-strategy

Contaminant Concentrations in Sport Fish from San Francisco Bay: 2019. Buzby, N.; Davis, J. A.; Sutton, R.; Miller, E.; Yee, D.; Wong, A.; Sigala, M.; Bonnema, A.; Heim, W.; Grace, R. 2021. SFEI Contribution No. 1036. San Francisco Estuary Institute: Richmond, CA. https://www.sfei.org/documents/contaminant-concentrations-sport-fish-san-francisco-bay-2019

Simulating Sediment Flux Through the Golden Gate. Prepared for Regional Monitoring Program for Water Quality in San Francisco Bay (RMP). Anchor QEA, L.L.C.. 2021. Simulating Sediment Flux Through the Golden Gate. Prepared for Regional Monitoring Program for Water Quality in San Francisco Bay (RMP). SFEI Contribution No. 1033. San Francisco Estuary Institute: Richmond, CA. https://www.sfei.org/documents/san-francisco-estuary-institute-collection-nist-biorepository

San Francisco Bay Regional Watershed Modeling Progress Report, Phase 1. Zi, T.; Mckee, L.; Yee, D.; Foley, M. 2021. SFEI Contribution No. 1038. San Francisco Estuary Institute: Richmond, CA. https://www.sfei.org/documents/san-francisco-bay-regional-watershed-modeling-progress-report-phase-1

The San Francisco Estuary Institute Collection at the NIST Biorepository. Ellisor, D.; Buzby, N.; Weaver, M.; Foley, M.; Pugh, R. 2021. NIST Interagency/Internal Report (NISTIR) - 8370. SFEI Contribution No. 1039. National Institute of Standards and Technology: Gaithersburg, MD. https://www.sfei.org/documents/san-francisco-estuary-institute-collection-nist-biorepository

PUBLISHED DATASETS

North Bay Bathymetry

High-resolution (1 m) digital elevation model (DEM) of San Francisco Bay, California, created using bathymetry data collected between 1999 and 2016. Fregoso, T.A., Jaffe, B.E., and Foxgrover, A.C., 2020. U.S. Geological Survey data release, https://doi.org/10.5066/P9TJTS8M

South Bay Suspended Sediment Concentration

Sediment concentration, water velocity, and suspended particle size and settling data to estimate estuarine sediment flux at Dumbarton Bridge, San Francisco Bay, CA from 2018 - 2019. Livsey, D.N., Downing-Kunz, M., and Einhell, D.C., 2020. U.S. Geological Survey data release, https://doi.org/10.5066/P9BVCW1N

Contaminant Concentrations in Sport Fish

Contaminant Concentrations in Sport Fish from San Francisco Bay: 2019. Buzby, N.; Davis, J. A.; Sutton, R.; Miller, E.; Yee, D.; Wong, A.; Sigala, M.; Bonnema, A.; Heim, W.; Grace, R. 2021. SFEI Contribution No. 1036. San Francisco Estuary Institute: Richmond, CA. https://www.sfei.org/documents/contaminant-concentrations-sport-fish-san-francisco-bay-2019

Collecting selenium samples in the North Bay. Photograph by Natasha Benjamin (Marine Applied Research and Exploration [MARE]). ▶

PROGRAM IMPACT

The **IMPACT** of the RMP on Management Decisions

Billions of dollars are at stake in decisions regarding activities that are directly intended to protect Bay water quality. The region has made huge investments to build and operate the infrastructure to collect and treat the region's sewage and industrial wastewater, and continued investment at a similar scale will be needed to maintain, upgrade, and operate this infrastructure to serve a growing Bay Area population. The region has spent and will continue to spend comparably large sums to manage contaminated soil and sediment in Bay watersheds, to manage stormwater, and to establish green infrastructure in our cities to capture stormwater and minimize its adverse water quality impacts on the Bay. Large investments have been and will be made to manage contaminated sediment in the Bay: at sites identified for cleanup, for dredging to maintain channels for commercial and recreational vessels, and for infrastructure to support using dredged sediment to restore wetlands and make the Bay shoreline more **resilient to rising sea level.**

Billions more are riding on decisions regarding activities that influence Bay water quality as unintentional side-effects. Commercial product formulation and usage (including pesticides, pharmaceuticals, personal care products, electrical equipment, home furnishings, automobile components, and many, many others), sediment management, water supply management, energy production, and habitat restoration and management are all immense and essential enterprises that have a tremendous influence on Bay water quality.

More than money is at stake. Protecting the health of people who eat fish and shellfish from the Bay is one of the primary objectives of water quality managers. Cleanup plans for many contaminants are driven by this objective, as are decisions regarding advisories to promote safe consumption of fish from the Bay. Cleanup plans also aim to protect the health of fish, wildlife, and all of the aquatic species that live in the Bay.

The goal of the RMP is to collect data and communicate information about Bay water quality in support of all of these management decisions. The $3.9 million annual budget for the RMP is used judiciously so that these decisions on Bay water quality are informed by sound science.

◀ Cargo ships in Central Bay.
Photograph by Ellen Plane.

Regulatory Policies Informed by the RMP

Management of pollutant discharges to the Bay: wastewater, stormwater, dredged material
Regional Water Board, US Environmental Protection Agency

303(d) Listings

Total Maximum Daily Load Control Plans (TMDLs)
- San Francisco Bay Mercury TMDL
- Guadalupe River Mercury TMDL
- San Francisco Bay PCBs TMDL
- North Bay Selenium TMDL
- Suisun Marsh TMDL for Dissolved Oxygen and Mercury

Permits
- National Pollutant Discharge Elimination System (NPDES) wastewater discharge permit provisions
- Municipal Regional Stormwater Permit - Load reductions, green infrastructure planning
- Mercury and PCBs Watershed Permit for Municipal and Industrial Wastewater
- Nutrient Watershed Permit for Municipal Wastewater

Criteria
- Site-specific objectives and implementation plans for copper and cyanide
- Nutrient numeric endpoint framework (under development)

Contaminant of Emerging Concern (CEC) Action Plans

Commercial product formulation and usage
California Department of Pesticide Regulation, Department of Toxic Substances Control, others

- State legislative bans: microbeads, PBDEs, copper in brake pads
- State flammability standards for furniture and building materials: flame retardants
- State pesticide regulations: e.g., pyrethroids
- State Safer Consumer Products regulations
- State product label changes: fipronil
- Federal legislative bans: PCBs, microbeads
- Federal pesticide regulations: DDT, chlordane, dieldrin, diazinon, and chlorpyrifos
- County and local drug take-back ordinances and programs

Dredging and dredged material management
US Army Corps of Engineers, San Francisco Regional Water Board, San Francisco Bay Conservation and Development Commission, US Environmental Protection Agency, and others

- Dredging and dredged material disposal permits through the Dredged Material Management Office
- Long-Term Management Strategy for the Placement of Dredged Material in the San Francisco Bay Region (LTMS)
- Essential Fish Habitat Agreement for Maintenance Dredging Conducted Under the LTMS Program
- Regional restoration plans

Public health protection
California Office of Environmental Health Hazard Assessment

- Fish consumption advice and communication

▼ Cargo ship on the Bay. Photograph by Marco Sigala.

RMP Impact Summary
Municipal Wastewater Dischargers

DECISIONS INFORMED BY THE RMP

- **Are treatment plant modifications or upgrades, or source reduction activities needed?**

  - **Which contaminants need to be reduced in municipal wastewater?**
    Examples of contaminants currently under consideration for reductions are nutrients, the pesticides fipronil and imidacloprid, and other contaminants of emerging concern.

  - **At which treatment plants are the reductions needed?**
    Different segments of the Bay vary greatly in their general characteristics, including in some cases their sensitivity to additional contaminant loads. The need for load reductions may therefore vary in different segments of the Bay.

  - **How much of a reduction is needed?**
    The goal of TMDLs and other control plans is to reduce concentrations in the Bay to levels that do not significantly impact beneficial uses. This requires a solid understanding of impairment and contaminant cycling in the Bay.

  - **What is the effect of the reductions or modifications on Bay water quality?**
    Monitoring is essential in demonstrating that load reduction efforts achieve the desired improvement in beneficial use attainment. Monitoring is needed to ensure that treatment plant modifications (e.g., implementation of reverse osmosis for water reuse) have no adverse impacts on beneficial uses.

- **Are actions needed for other pathways to reduce loads and impairment from contaminants found in municipal wastewater?** A holistic understanding of the relative importance of loads for all pathways is needed to optimize overall load reduction efforts.

REGULATIONS ADDRESSED

NPDES Permits

Mercury TMDL

PCBs TMDL

North Bay Selenium TMDL

Copper Site-Specific Objective (SSO) Implementation Plan

Nutrient Watershed Permit

Mercury and PCBs Watershed Permit

CEC Action Plans

Cyanide SSO Implementation Plan

Department of Toxic Substances Control (DTSC) Safer Consumer Product Regulations

Department of Pesticide Regulation (DPR) state pesticide regulations

USEPA Federal Insecticide, Fungicide, and Rodenticide Act

RMP Impact Summary
Municipal Stormwater Dischargers

DECISIONS INFORMED BY THE RMP

- **Which contaminants need to be reduced in municipal stormwater?** Reductions of legacy contaminants are currently a primary focus of stormwater management attention, but other contaminants, including contaminants of emerging concern, may also need to be reduced.

- **How much load reduction effort is needed?** The goal of TMDLs and other control plans is to reduce concentrations in the Bay to levels that do not significantly impact beneficial uses. This requires a solid understanding of the linkage between stormwater and Bay impairment.

- **Which tributaries should be priorities for actions to reduce loads?** Different segments of the Bay encompass variable watershed source areas and related loads, and vary greatly in their general characteristics, including in some cases their sensitivity to additional contaminant loads. The need for load reductions may therefore vary for tributaries discharging to different segments of the Bay.

- **Which sources or source areas in watersheds should be targeted for load reductions?** Identifying the sources and source areas in watersheds to target is a major challenge in reducing stormwater loads.

- **What is the effect of load reductions or other stormwater management and watershed modifications on Bay water quality?** Monitoring and modeling are essential to demonstrating that load reduction efforts achieve the desired improvement in beneficial use attainment. Other activities in the watershed (e.g., land use changes or changes in chemical use) may also affect contaminant loads in either beneficial or adverse ways.

- **Are actions needed for other pathways to reduce loads and impairment from contaminants found in municipal stormwater?** A holistic understanding of the relative importance of loads for all pathways is needed to optimize overall load reduction efforts.

REGULATIONS ADDRESSED

NPDES Permits

Municipal Regional Stormwater Permit

Mercury TMDL

PCBs TMDL

North Bay Selenium TMDL

Copper Site-Specific Objective Implementation Plan

CEC Action Plans

DTSC Safer Consumer Product Regulations

DPR state pesticide regulations

USEPA Federal Insecticide, Fungicide, and Rodenticide Act

RMP Impact Summary
Industrial Wastewater Dischargers

DECISIONS INFORMED BY THE RMP

- **Are treatment plant modifications or upgrades, or source reduction activities needed?**

  - **Which contaminants need to be reduced in industrial wastewater?** For example, the need for selenium reductions in refinery effluent was identified in the 1990s, and treatment upgrades implemented in the late 1990s achieved large reductions in selenium loads.

  - **At which treatment plants are the reductions needed?** Specific industrial discharges may contain higher levels of chemicals that may merit special attention. For example, sites where fire-fighting foams have been used may discharge higher levels of PFOS, a chemical of emerging concern present in older formulations. In addition, different parts of the Bay vary greatly in their general characteristics, including in some cases their sensitivity to additional contaminant loads. The need for load reductions may therefore vary in different segments of the Bay.

  - **How much of a reduction is needed?** The goal of TMDLs and other control plans is to reduce concentrations in the Bay to levels that do not significantly impact beneficial uses. This requires a solid understanding of impairment and contaminant cycling in the Bay.

  - **What is the effect of the reductions or modifications on Bay water quality?** Monitoring is essential in demonstrating that load reduction efforts achieve the desired improvement in beneficial use attainment. Monitoring is needed to ensure that treatment plant modifications (e.g., implementation of reverse osmosis for water reuse) have no adverse impacts on beneficial uses.

- **Are actions needed for other pathways to reduce loads and impairment from contaminants found in industrial wastewater?** A holistic understanding of the relative importance of loads for all pathways is needed to optimize overall load reduction efforts.

REGULATIONS ADDRESSED

NPDES Permits

Mercury TMDL

PCBs TMDL

North Bay Selenium TMDL

Copper SSO Implementation Plan

Mercury and PCBs Watershed Permit

CEC Action Plans

DTSC Safer Consumer Product Regulations

RMP Impact Summary
Dredgers

DECISIONS INFORMED BY THE RMP

- **Where can contaminated dredged material be disposed?** RMP sediment data are the basis for the Dredged Material Testing Thresholds for mercury, polycyclic aromatic hydrocarbons (PAHs), and PCBs. These thresholds determine when bioaccumulation testing is required for dredged material to be discharged at unconfined open water disposal sites in the Bay. RMP sediment data also serve as the basis for in-Bay dredged material disposal limits called for in the PCBs and mercury TMDLs.

- **Should dredged material be reused within the Bay and where?** Management of sediment as a resource in the Bay requires understanding of the volumes, types, locations, and environmental drivers of sediment input. The RMP performs extensive monitoring of suspended sediment concentrations along with monitoring of suspended sediment loads at select tributaries. The RMP also funds special studies to understand sediment transport within the Bay.

- **Should dredging practices be modified to prevent impacts to fish and benthic species?** The benthic communities of the Bay provide important foraging habitat for many fish species. The RMP performs studies to understand whether dredging practices have an impact on benthic species and habitats. The RMP also studies whether exposure to contaminants in dredged material poses a risk to fish.

REGULATIONS ADDRESSED

2011 Programmatic Essential Fish Habitat Agreement, Measure 1

2011 Programmatic Essential Fish Habitat Agreement, Measure 7

PCBs TMDL

Mercury TMDL

Long-Term Management Strategy

PROGRAM AREA
UPDATES

BACKGROUND

The Status and Trends monitoring program is the core of the RMP's long-term monitoring strategy. Since the beginning of the RMP in 1993, water, sediment, and bivalve tissue have been monitored regularly in the open Bay. Sport fish and bird egg monitoring were added to the Program in 1997 and 2007, respectively.

Annual sampling of water and sediment had sufficiently documented trends and spatial patterns that varied by pollutant. This led to a reduction in frequency between 2011 and 2014 to free up resources for special studies and other topics.

Sediment monitoring in the shallow margin areas of the Bay is currently being considered for addition to the Status and Trends program. Pilot studies were completed in Central, South, and North Bays in 2015, 2017, and 2020, respectively.

The Status and Trends Review will be complete at the end of 2021 and implemented according to the revised sampling schedule.

USES OF PROGRAM AREA DATA FOR MANAGEMENT DECISIONS

- Defining ambient conditions in the Bay
- Water Quality Assessment — 303(d) impairment listings or de-listings
- Determination of whether there is reasonable potential that a NPDES-permitted discharge may cause violation of a water quality standard
- Evaluation of water and sediment quality objectives

- Dredged material management
- Development and implementation of TMDLs for mercury, PCBs, and selenium
- Site-specific objectives and antidegradation policies for copper and cyanide
- Development and evaluation of a Nutrient Assessment Framework (i.e., development of water quality objectives)

RELATION TO PERMIT REQUIREMENTS

NPDES Permits

- Receiving water compliance monitoring for NPDES discharge permit holders
- Provides data for Reasonable Potential Analyses
- Provides data for evaluating site specific objectives for copper and cyanide

Essential Fisheries Habitat Consultation, PCBs TMDL, Mercury TMDL

- Provides data to calculate ambient dredged material testing guidelines

PRIORITY QUESTIONS

1 What are the concentrations and masses of priority contaminants in the Bay, its compartments, and its segments?

2 Are contaminants at levels of concern?

3 Are there particular regions of concern?

4 Have concentrations and masses increased or decreased?

5 What are the associated impacts of those contaminants?

RECENT FINDINGS

In 2019, the RMP monitored five species of sport fish at thirteen locations throughout the Bay (Featured Project on page XX). Mercury and PCB concentrations remain above water quality objectives and consumption thresholds, while dioxin concentrations remain above a Water Board screening level. Selenium and PBDE concentrations were below levels of concern for human health.

Preliminary analysis of the North Bay margins sediment collected in 2020 suggest that concentrations in the margins are mostly lower than those in the margins of South and Central Bay, which have higher densities of urban and industrial development.

Review of the current S&T Program continued through 2021. The water design was finalized in January; the sediment design in June; and the biota design in August. A synthesis meeting in September ensured that all three designs are well aligned, are providing the necessary data to inform management decisions, and can be completed with available funding.

WORKPLAN HIGHLIGHTS

The S&T workplan is currently under review and the timing of long-term monitoring is likely to shift starting in 2022. The current frequency of sampling includes:

- nutrients monthly,
- water every two years,
- bivalves every two years (discontinued in 2020; sampling at locations around the edge of the Bay will be assessed during the S&T Review),
- bird eggs every three years,
- sediment once every four years, and
- sport fish once every five years.

In 2020, the RMP monitored sediment in the North Bay margins. A report summarizing the results of all three margins sampling efforts will be completed in early 2022.

In 2021, the RMP is conducting biennial Bay water monitoring. The legacy pollutant analyte list for this cruise is reduced and will focus on copper based on the outcomes of the Status & Trends Review. In addition, two emerging contaminants are being sampled: bisphenols and organophosphate esters (samples for PFAS will also be collected as a Special Study). Bird egg sampling in 2021 was pushed out to 2022 due to COVID-related fieldwork issues. The RMP continues to collaborate with the US Geological Survey on the fortnightly South Bay and monthly Bay-wide cruises to assess nutrient and phytoplankton conditions in the Bay. The northern extent of the cruises have been limited to San Pablo Bay due to COVID, but all stations throughout Lower South, South, and Central Bays are being collected.

COLLABORATORS

- San Francisco Bay Regional Water Quality Control Board
- US Environmental Protection Agency
- Applied Marine Sciences
- SGS AXYS
- Brooks Analytical Labs
- Eurofins Scientific
- Caltest Analytical Laboratory
- San Francisco Public Utilities Commission Water Quality Division
- US Geological Survey
- ALS Environmental
- Pacific EcoRisk
- Moss Landing Marine Laboratory
- Marine Pollution Studies Laboratory
- Coastal Conservation & Research
- City of San Jose

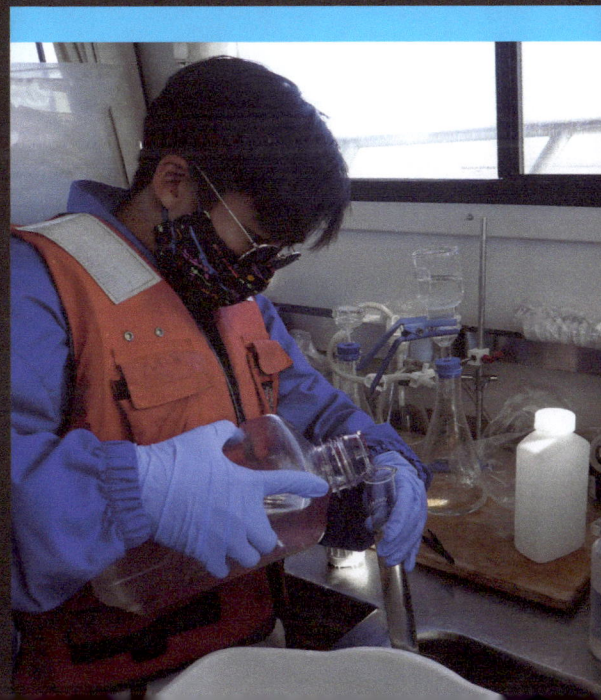

| # EMERGING CONTAMINANTS

BACKGROUND

Contaminants of emerging concern (CECs) are generally not actively regulated or routinely monitored, yet have the potential to enter the environment and harm people or aquatic life. Through its focus on CECs, the RMP aims to identify problem chemicals before they cause harm. The RMP's decades-long effort has made the Bay one of the most thoroughly studied estuaries in the world for CECs. Surveillance has identified several contaminants or contaminant classes of moderate concern:

- PFAS — stain and water repelling chemicals widely used in industrial and consumer products;

- fipronil and imidacloprid — insecticides with widespread urban uses;

- alkylphenols and alkylphenol ethoxylates — detergent ingredients;

- bisphenols — plastic additives; and

- organophosphate esters — flame retardants and plasticizers.

PRIORITY QUESTIONS

1 Which CECs have the potential to adversely impact beneficial uses in San Francisco Bay?

2 What are the sources, pathways, and loadings leading to the presence of individual CECs or groups of CECs in the Bay?

3 What are the physical, chemical, and biological processes that may affect the transport and fate of individual CECs or groups of CECs in the Bay?

4 Have the concentrations of individual CECs or groups of CECs increased or decreased in the Bay?

5 Are the concentrations of individual CECs or groups of CECs predicted to increase or decrease in the future?

6 What are the effects of management actions?

RELATION TO PERMIT REQUIREMENTS

- Municipal wastewater dischargers may opt into the alternate monitoring permit requirements with fees that provide additional funds to support the RMP and its CEC monitoring.

- The most recent Municipal Regional Stormwater Permit (2015) requires monitoring studies of key CECs, including flame retardants, PFAS, and pesticides.

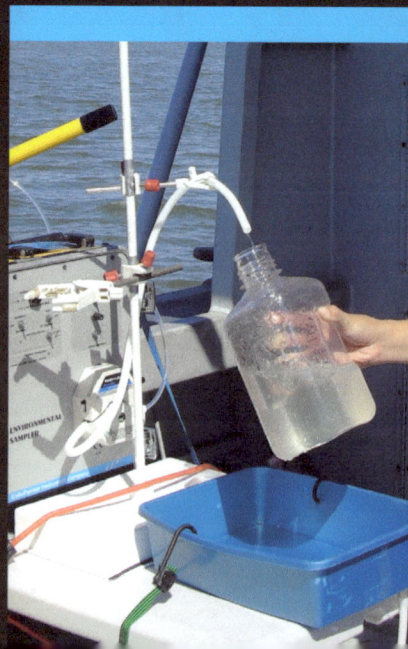

USES OF PROGRAM AREA DATA FOR MANAGEMENT DECISIONS

- Regional Action Plans for CECs

- Early management intervention, including green chemistry and pollution prevention

- State and federal pesticide regulatory programs

- State Water Board CEC Initiative

- DTSC Safer Consumer Products Program

RECENT FINDINGS

In December 2020, RMP collaborators in the State of Washington announced the findings of a decades-long effort to identify the cause of coho salmon deaths in Puget Sound streams. The contaminant, 6PPDQ, is derived from a tire preservative (6PPD), and can wash into streams along with tire wear particles when it rains.

RMP scientists collected samples from nine Bay Area streams and storm drains during storm events as part of an ongoing multi-year study to screen urban stormwater runoff for CECs; four samples contained levels of 6PPDQ above the concentration at which half the coho salmon die after a few hours of exposure in laboratory experiments. Coho salmon no longer reside in San Francisco Bay and its streams, but they are being restored to coastal streams from Santa Cruz to Sonoma County. They currently populate the Klamath, Smith, and Eel Rivers further north. Researchers are also concerned that steelhead trout and Chinook salmon exhibit some sensitivity to tire rubber chemicals, and studies are ongoing on those species.

The San Francisco Bay Microplastics Project, a study completed in 2019 and supported in part by the RMP, found that nearly half of the estimated seven trillion microplastic particles in urban stormwater flowing through local streams into the Bay could potentially be linked to tire wear. The 6PPDQ study, published in the journal Science, indicates such particles can be toxicologically relevant.

In response to these important findings, as well as a petition from the state's stormwater leaders to act on zinc (another harmful chemical in tires), California's Department of Toxic Substances Control is considering taking action on motor vehicle tires containing zinc and 6PPD. To support decision-making, SFEI's Kelly Moran and Rebecca Sutton presented RMP science as part of a July 2021 public workshop on the chemicals in tires.

In addition, RMP monitoring continues on classes of CECs considered to be moderate concerns for the Bay. This spring, the RMP reviewed recent findings on alcohol and alkylphenol ethoxylated surfactants in Bay Area stormwater runoff, wastewater effluent, and ambient Bay water. These ethoxylated surfactants are high production volume chemicals commonly used in industrial and consumer applications as detergents and emulsifiers in paints, cleaning products, personal care products, pesticides, and in the textile, paper, and metal industries. Ethoxylated surfactants were widely observed in both wastewater and stormwater. While total concentrations were generally similar for the pathways, a few effluent samples contained unusually high levels, with concentrations up to 45 µg/L. Levels in Bay water were generally low, with significant detections at only two sites.

COLLABORATORS

- Bay Area Clean Water Agencies
- California Department of Toxic Substances Control
- California Department of Pesticide Regulation
- Colorado School of Mines
- Duke University
- Jinan University
- SGS AXYS
- San Francisco Bay Regional Water Quality Control Board
- San Diego State University
- Southern Illinois University
- Southern California Coastal Water Research Project
- University of Minnesota
- University of Washington

WORKPLAN HIGHLIGHTS

Multi-year Monitoring Effort for CECs in Stormwater Nearing Completion: Findings from RMP non-targeted analysis resulted in a new focus on unique and rarely studied contaminants derived from vehicles and roadways, such as 6PPDQ. A major effort to investigate these and other CECs in Bay Area stormwater began in 2019 and continues with a final year of sample collection, starting this fall. In parallel with monitoring, we will begin developing an overall strategy for monitoring CECs in stormwater.

Tire and Roadway Contaminants in Dry and Wet Season Bay Water: 6PPDQ and other toxicologically relevant contaminants derived from tires and vehicles have been observed in stormwater, but have not yet been quantified in the Bay. The RMP will determine levels of these contaminants in the Lower South Bay in both the dry and wet season, and will also analyze Bay water samples collected adjacent to stormwater discharge points right after storms. Findings will inform placement of these contaminants within the RMP's tiered risk-based framework for prioritizing CECs.

Expanded Monitoring of Ethoxylated Surfactants in Bay Water, Margin Sediment, and Wastewater: A follow-up study will support method development and reanalysis of archived samples to include nonylphenol, octylphenol, and short chain ethoxylates in addition to the analytes quantified previously. Additional wastewater samples will be collected to confirm the range of values quantified previously. The full dataset will guide development of a monitoring strategy for these contaminants.

SMALL TRIBUTARY LOADING

BACKGROUND

Pollutants of concern (POCs) in urban stormwater include PCBs, mercury, copper, nutrients, pesticides, contaminants of emerging concern (CECs), and microplastics. To address information needs associated with these POCs, the Small Tributaries Loading Strategy (STLS), first developed in 2009, was updated in 2018 to include a modeling and trends component to help prioritize and coordinate the activities of the RMP and Bay Area Municipal Stormwater Collaborative permittees. STLS studies conducted over the past decade have focused on locating, quantifying, and managing PCBs, mercury, and other pollutants in the urban environment to support management actions.

Going forward, an increasing emphasis will be placed on CECs, along with tracking trends in POC loading, through a combination of monitoring and modeling. Since these emphases were not considered at the time of the last management question update (2015), in May 2021 the Sources, Pathways, and Loadings Workgroup (SPLWG) began discussing the need for a formal update of the management questions that guide the SPLWG.

USES OF PROGRAM AREA DATA FOR MANAGEMENT DECISIONS

- Refining pollutant loading estimates, including CECs, for future policy or management plan updates (collaboration with ECWG)

- Informing provisions of the current and future versions of the Municipal Regional Stormwater Permit (MRP)

- Identifying small tributaries to prioritize for management actions

- Informing decisions on the control measures for reducing pollutant concentrations and loads

- Tracking effectiveness of load reduction in small tributaries

- Estimating sediment loads to the Bay (collaboration with Sediment Workgroup)

RELATION TO PERMIT REQUIREMENTS

Addresses monitoring requirements specified in the Municipal Regional Stormwater Permit

- POC and CEC monitoring

- Wet weather pesticide and toxicity monitoring

- Implementation of control measures to achieve mercury and PCB load reductions

- Assessment of mercury and PCB load reductions from stormwater

- Planning and implementation of green infrastructure to reduce mercury and PCB loads

- Preparation of implementation plan and schedule to achieve TMDL allocations

PRIORITY QUESTIONS

1 What are the loads or concentrations of pollutants of concern from small tributaries to the Bay?

2 Which are the "high-leverage" small tributaries that contribute or potentially contribute most to Bay impairment by pollutants of concern?

3 How are loads or concentrations of pollutants of concern from small tributaries changing on a decadal scale?

4 Which sources or watershed source areas provide the greatest opportunities for reductions of pollutants of concern in urban stormwater runoff?

5 What are the measured and projected impacts of management action(s) on loads or concentrations of pollutants of concern from the small tributaries, and what management action(s) should be implemented in the region to have the greatest impact?

Note: Recent workgroup discussion pointed to the fact that some of these management questions may need revising in relation to changing emphases and greater cross-workgroup collaboration.

RECENT FINDINGS

Winter storm sampling by the RMP, utilizing a combination of manual sampling as well as unattended remote suspended sediment sampling, has been conducted in 91 watersheds. The watersheds with the highest PCB concentrations on exported particles are Pulgas Creek Pump Station in San Carlos, a ditch on Industrial Road in San Carlos, Line 12H at Coliseum Way in Oakland and Santa Fe Channel in Richmond. Outfalls at Gilman Street and the Santa Fe Channel sites also appear to have relatively high concentrations of mercury. Because water years 2020 and 2021 were notable for being the second driest consecutive years on record in the Bay Area, information has not increased greatly in the past two years.

To make better use of data from watersheds exhibiting moderate or lower concentrations, in 2018 the RMP began exploring new data analysis methods based on loads and yields and PCB congener patterns to provide additional information to support management decisions. With method development completed in 2018, the loads and yields methods have now been applied to over 130 watersheds and the congener analysis method was applied in 75 watersheds (limited due to data availability). This analysis revealed some new insights on areas within watersheds to consider for PCB management and also provided corroborating evidence for other watersheds where management is already underway.

In 2018, a Regional Watershed Spreadsheet Model (RWSM) was completed with a reasonable calibration for hydrology, PCBs, and Hg for annual average simulation. This model has so far provided support for reasonable assurance analysis and other planning efforts for PCBs and mercury, and estimates for regional-scale trash assessments, copper, and microplastics. Workgroup advisors have recommended the continued use of this model for making regional-scale loads estimates for pollutants with scarce data.

In 2020, development began on a new dynamic regional watershed model (WDM) for Bay Area hydrology, sediment, and pollutant loads and trends. The spatial domain of this new model is the area that drains to San Francisco Bay from the nine adjacent counties around the Bay. The Loading Simulation Program in C++ (LSPC) modeling framework was selected to simulate large complex regions with mixed land use types, a wide range of pollutants, upland erosion and sediment transport, and in-stream processes (e.g., bank erosion, settling, and resuspension). The completed hydrologic model is performing well, reproducing the timing and peaks of runoff events as well as the annual and intra-annual variation of hydrological processes.

WORKPLAN HIGHLIGHTS

Stormwater Reconnaissance Sampling: Over the past seven years, the RMP, in collaboration with county stormwater programs, has funded watershed reconnaissance for the identification and management of PCB and mercury sources. This effort is continuing, providing data on concentrations in water and on sediment particles to identify high-leverage watersheds and subwatersheds within larger areas of older urban and industrial land use. Additionally, monitoring site selection will include prioritization for CECs to support Emerging Contaminant Workgroup efforts to characterize watershed concentrations, and considerations to support calibration and verification of the PCB and mercury dynamic regional models at sites where flow monitoring is occurring.

Regional Loads Estimates by RWSM: In 2022, there is a small project to update the RWSM to a more recent climatic period (1991-2020), update the land use dataset, make other minor code improvements (including considering conceptual models for CECs as a basis for structural model changes), and recalibrate the model. In parallel, another small project in 2022 will explore the potential of using the RWSM or other methods to make estimates of loads for CECs.

Trends Strategy and Watershed Dynamic Model (WDM) Development: The evaluation of stormwater loading trends in relation to management efforts and beneficial use impacts is an important new focus. The hydrology module of the WDM was calibrated and completed in 2020 and serves as a solid foundation for the sediment (2021) and future pollutant model development and simulations (2022 and beyond). The WDM is a flexible and powerful tool that will be used for synthesizing understanding of regional hydrologic, sediment, and POC (both legacy and emerging) loading processes and linking those to processes in the Bay through a new pilot study beginning in 2022 that will develop a linked watershed-Bay model for the first time. Although the medium-term plan is for the model to broadly support a wide range of POCs, including CECs, the first POC simulation effort will be applied to pollutants with the most data, PCBs and mercury.

Integrated Watershed Modeling and Monitoring Implementation Strategy: As the focus of watershed modeling moves towards assessing a broader suite of contaminants, including CECs, sediment, and nutrients, the monitoring required to model contaminant groups with similar characteristics (e.g., chemical and physical properties, sources, pathways) needs to be systematically identified. A RMP integrated watershed modeling and monitoring strategy (2021) is being developed in coordination with the ECWG and will address management questions related to watershed loading of PCBs, mercury, CECs, and other constituents. In addition, the ECWG will be working on a stormwater monitoring strategy in 2022 while the SPLWG will be working on a stormwater modeling strategy for CECs. These three new products will inform an integrated watershed monitoring and modeling approach in 2022 and beyond and will support the development of the WDM in 2022.

COLLABORATORS

- Bay Area Municipal Stormwater Collaborative
- San Francisco Bay Regional Water Quality Control Board
- US Geological Survey
- California Department of Pesticide Regulation

PROGRAM AREA UPDATE | NUTRIENTS

BACKGROUND

San Francisco Bay receives some of the highest nitrogen loads among estuaries worldwide, yet has not historically experienced the water quality problems typical of other nutrient-enriched estuaries. It is not known whether this level of nitrogen loading, which will continue to rise in proportion to human population increase, is sustainable over the long term. Special studies and expanded monitoring carried out through the RMP and the Nutrient Management Strategy have revealed some water quality conditions that have been associated with nutrient over-enrichment in other estuaries (e.g., recurring low dissolved oxygen in some margin habitats and consistent detection of multiple toxins produced by harmful algae). Potential impacts of these conditions on human and ecological health need to be more extensively evaluated and causal factors determined. A further complication is that the Bay's response to nutrients is influenced by many physical and biological factors including suspended sediment concentrations, light availability, freshwater inputs, and ocean conditions. These factors themselves vary by Bay subembayment and due to regional land and water management and climate oscillations. Therefore, a wide range of monitoring and special studies is needed to understand what might happen to Bay water quality as a result of changes in nutrients and other factors.

USES OF PROGRAM AREA DATA FOR MANAGEMENT DECISIONS

- Developing nutrient numeric endpoints and an assessment framework
- Evaluating the need for revised objectives for dissolved oxygen and other parameters
- Assessing water quality impairment status
- Implementing NPDES permits for wastewater and stormwater

RELATION TO PERMIT REQUIREMENTS

The Bay-wide nutrient permit for municipal wastewater that went into effect in 2014 includes a provision to support science and monitoring to inform future permitting decisions. The second five-year Bay-wide nutrient permit started in 2019.

PRIORITY QUESTIONS

1 What conditions in different Bay habitats would indicate that beneficial uses are being protected versus experiencing nutrient-related impairment?

2 In which subembayments or habitats are beneficial uses being supported? Which subembayments or habitats are experiencing nutrient-related impairment?

3 To what extent is nutrient over-enrichment, versus other factors, responsible for current impairments?

4 What management actions would be required to mitigate those impairments and protect beneficial uses?

5 Under what future scenarios could nutrient-related impairments occur, and which of these scenarios warrant pre-emptive management actions?

6 What management actions would be required to protect beneficial uses under those scenarios?

7 What nutrient sources contribute to elevated nutrient concentrations in subembayments or habitats that are currently impaired, or would be impaired in the future, by nutrients?

8 When nutrients exit the Bay through the Golden Gate, where are they transported and how do they influence water quality in the Gulf of Farallones or other coastal areas?

9 What specific management actions, including load reductions, are needed to mitigate or prevent current or future impairment?

RECENT FINDINGS

Nutrient loads to San Francisco Bay are increasing. Combined nitrogen loads from the region's five largest wastewater treatment plants increased 25-30% between 2000 and 2018.

High-frequency sensors are providing continuous data at nine sites in South Bay and Lower South Bay. These data show that elevated phytoplankton biomass and low dissolved oxygen are frequently observed in Lower South Bay margin habitats, and suggest that water from the salt ponds introduces high phytoplankton biomass into Lower South Bay sloughs and increases the potential for low dissolved oxygen events. Unprecedented smoke from wildfires led to the lowest dissolved oxygen concentrations ever observed by the NMS in the Lower South Bay. The absence of light resulted in a shift in the metabolic balance of the system, causing oxygen concentrations to plummet and putting fish and other biota at risk.

Harmful algal bloom-forming phytoplankton species are commonly detected throughout the Bay, and multiple HAB toxins occur in water samples, anchovies, and mussels.

Current estimates suggest that San Francisco Bay is a significant source of nutrients to the coastal ocean. Ongoing modeling work aims to identify how this input affects coastal condition.

Progress continues on model simulations of nutrient transport, phytoplankton blooms, oxygen cycling, nutrient transformations, and other processes.

WORKPLAN HIGHLIGHTS

- Conducting experiments in South Bay to measure biogeochemical transformation rates
- Collecting data in Lower South Bay to assess the link between salt pond production and sloughs
- Determining healthy DO-related habitat conditions in Lower South Bay sloughs and creeks and other margin habitats
- Forecasting conditions in the Bay under potential future scenarios
- Assessing the fate of nutrients that leave San Francisco Bay and the effects along the coast
- Investigating the mechanistic link between nutrients and harmful algal toxins and blooms
- Developing trends analyses for key indicators of water quality
- Refining indicators and metrics that are included in the Assessment Framework for the deep subtidal areas of the Bay
- Expanding high-frequency monitoring on the dynamic shoals of South Bay
- Linking nutrient transformation rates measured in the field to the development of the biogeochemical model.

COLLABORATORS

- San Francisco Bay Regional Water Quality Control Board
- Bay Area Clean Water Agencies
- Deltares
- San Francisco State University
- Stanford University
- UC Berkeley
- UC Davis

- UC Santa Cruz
- US Environmental Protection Agency
- US Geological Survey – Sacramento
- US Geological Survey – Menlo Park
- US Geological Survey - Santa Cruz
- University of Maryland Center for Environmental Science

PROGRAM AREA
UPDATE | PCBs

BACKGROUND

PCB contamination is a high priority for Bay water quality managers due to concerns for risks to humans and wildlife. A TMDL was approved in 2009, but concentrations in Bay sport fish have not declined since then, or even since RMP sport fish monitoring began in 1997.

In 2014, the RMP completed a synthesis report summarizing advances in understanding of PCBs in the Bay since the data synthesis for the PCBs TMDL. An updated conceptual model presented in that report called for monitoring and management to focus on contaminated areas on the Bay margins. Local-scale actions within margin areas, or in upstream watersheds, will be needed to reduce exposure within these areas. The multi-year workplan for PCBs is focusing on supporting a possible revision of the PCBs TMDL by evaluating the likelihood of improvements in high-priority margin areas in response to anticipated stormwater load reductions, and by establishing baselines for monitoring these improvements.

Site-specific conceptual models have been developed for three margin areas that are high priorities for water quality managers: the Emeryville Crescent, San Leandro Bay, and Steinberger Slough/Redwood Creek. Planning is underway for development of a mechanistic fate model for PCBs and other contaminants in the Bay.

USES OF PROGRAM AREA DATA FOR MANAGEMENT DECISIONS

- PCBs TMDL – support for appropriate changes to the TMDL

- NPDES Municipal Regional Stormwater Permit and wastewater permit requirements

- Focusing management actions and/or locations for reducing PCB impairment (upland)

- Determining cleanup priorities (in-Bay)

- Updating the fish consumption advisory

RELATION TO PERMIT REQUIREMENTS

- Addresses critical information needs identified in the PCB TMDL related to municipal and industrial wastewater dischargers and stormwater management agencies

- Addresses a requirement in the Municipal Regional Stormwater Permit: Fate and transport study of PCBs - Urban runoff impact on San Francisco Bay margins

PRIORITY QUESTIONS

1 What are the rates of recovery of the Bay, its segments, and in-Bay contaminated sites from PCB contamination?

1a What would be the impact of focused management of PMU watersheds?

1b What would be the impact of management of in-Bay contaminated sites (e.g., removing and/or capping hot spots), both within the sites and at a regional scale?

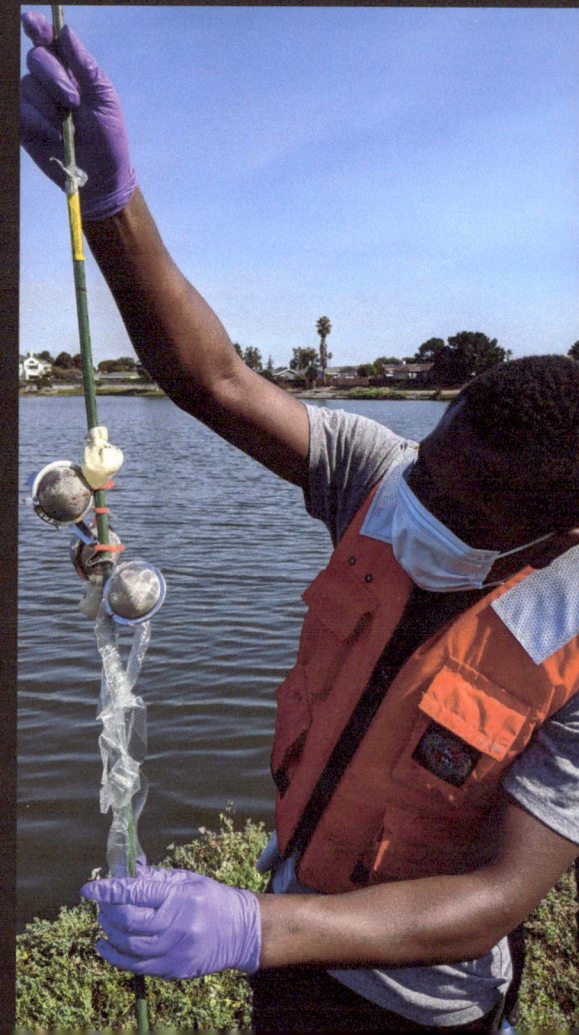

RECENT FINDINGS

In 2019, shiner surfperch had a Bay-wide average concentration 18 times higher than the TMDL target. These concentrations have resulted in an advisory from the Office of Environmental Health Hazard Assessment (OEHHA) recommending no consumption for all surfperch in the Bay. Concentrations in shiner surfperch and white croaker show limited signs of decline.

Urban stormwater is the pathway carrying the greatest PCB loads to the Bay and with the greatest load reduction goals. Concentrations of PCBs and mercury on suspended sediment particles from a wide range of watersheds have been measured as an index of the degree of watershed contamination and potential for effective management action. The three sites with the highest estimated particle PCB concentrations as of 2019 were Pulgas Pump Station South (8,220 ng/g), Industrial Rd Ditch in San Carlos (6,139 ng/g), and Line 12H at Coliseum Way in Oakland (2,601 ng/g).

Assessments of three "priority margin units" (the Emeryville Crescent, San Leandro Bay [SLB], and the Steinberger Slough/Redwood Creek area [SS/RC]) established conceptual models as a foundation for monitoring response to load reductions and for planning management actions. A key finding was that PCB concentrations in sediment and the food webs in the Crescent and SLB could potentially decline fairly quickly (within 10 years) in response to load reductions from the watershed. In contrast, recovery in SS/RC appears likely to be limited ultimately limited by the relatively high PCB concentrations that prevail in the South Bay segment of the Bay at the regional scale.

In spite of the expected responsiveness of SLB, extensive field studies there have documented persistent sediment contamination that is likely due to continuing inputs from the watershed.

WORKPLAN HIGHLIGHTS

- Baseline monitoring of four priority margin areas (Emeryville Crescent, San Leandro Bay, Steinberger Slough, and Richmond Harbor), including shiner surfperch monitoring in 2019 (in coordination with the 2019 Status and Trends sport fish monitoring).

- Field studies to address critical information gaps and establish baselines for evaluating the effects of load reductions in Steinberger Slough/Redwood Creek (2020, 2022) and San Leandro Bay (2021).

- Writing a plan in 2021 and beginning in 2022 to develop a model to forecast the fate of PCBs and other contaminants in the Bay, leveraging and integrating with models for nutrients, sediment, and watershed contaminant loads.

PARTNERS

- Moss Landing Marine Laboratory

- SGS AXYS Analytical

- Stanford University

- Integral Consulting Inc.

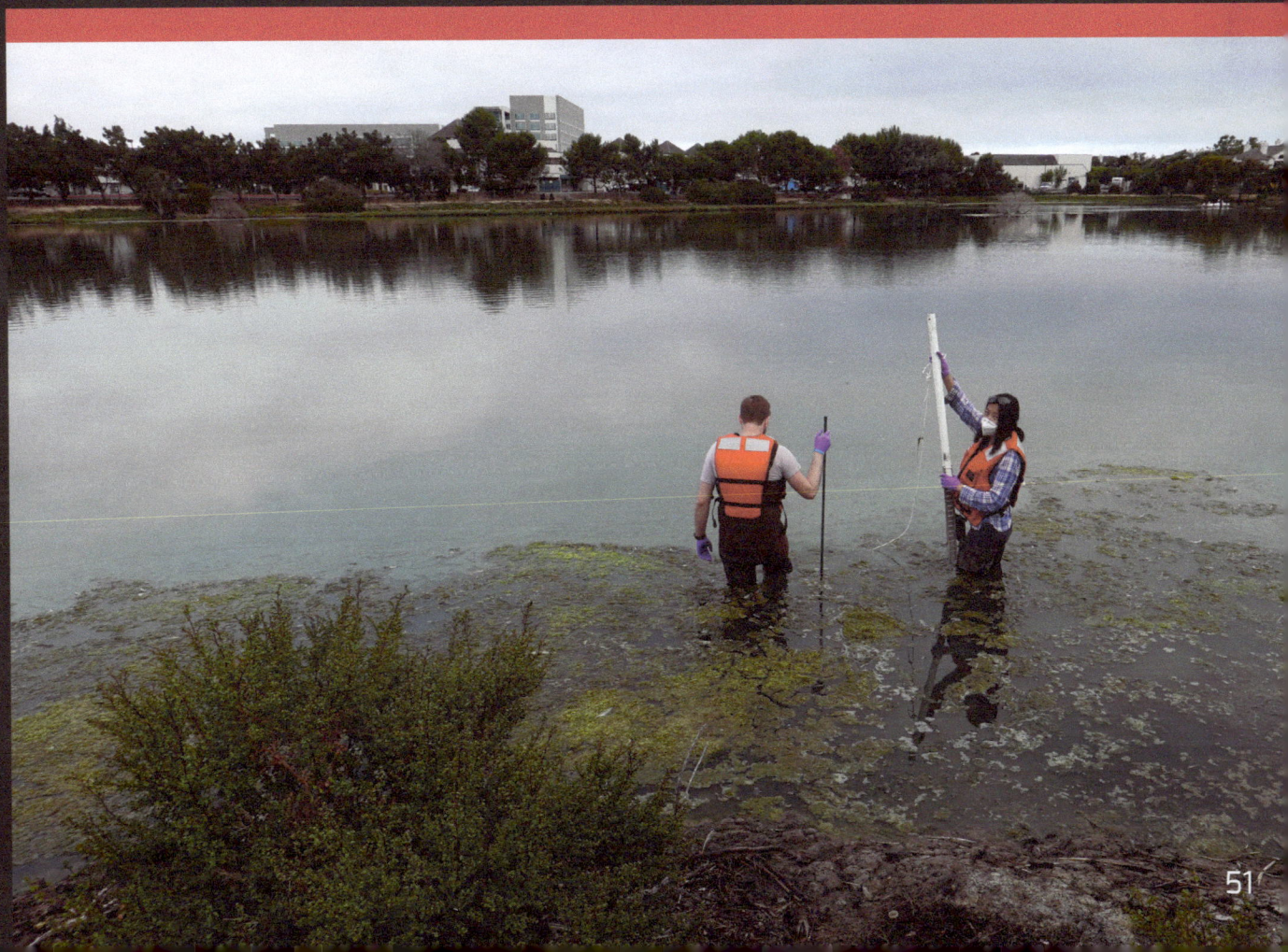

MICROPLASTICS

BACKGROUND

Microplastics, commonly defined as plastic particles smaller than 5 mm, come in a broad range of polymer types, shapes, and sizes. These properties affect the way microplastic particles move through the environment, and may modify their potential for toxicity. Information on the chemistry and morphology of particles can help to identify sources and options to mitigate the impact. While microplastics are abundant and ubiquitous, there is limited understanding of the ecological and human health risks related to microplastics. Recent legislation requires the California Ocean Protection Council to develop a state-wide microplastics strategy that articulates the risks from microplastics in marine waters and develop a plan for mitigation.

The San Francisco Estuary Institute recently completed a pioneering, three-year comprehensive regional study of microplastic pollution of a major urban estuary and adjacent ocean environment. This $1 million effort was primarily funded by the Gordon and Betty Moore Foundation, with additional funding and support provided by the RMP. Collaborators included science and advocacy organization 5 Gyres as well as scientists with the University of Toronto and University of California at Davis.

Findings were released at a one-day symposium, Science and Solutions for Microplastic Pollution, in October 2019. The symposium provided a summary of the state of the science in the morning. Afterwards, keynote speaker Jared Blumenfeld, Secretary for Environmental Protection, CalEPA, kicked off a dynamic discussion of potential solutions and actions to address microplastic pollution. Major project deliverables included a 400 page report on the scientific findings, a document outlining policy recommendations and solutions, an action sheet for broad public distribution, and a short documentary film. The findings received significant media coverage and multiple journal articles have been published.

USES OF PROGRAM AREA DATA FOR MANAGEMENT DECISIONS

- State-wide microplastic strategy
- State-wide drinking water monitoring
- Regional or state bans on single use plastic items and foam packaging materials
- State and federal bans on microbeads
- Statewide trash requirements
- Municipal pollution prevention strategy, including uses of green stormwater infrastructure
- Public outreach and education regarding pollution prevention

RELATION TO PERMIT REQUIREMENTS

There are no current permit requirements for microplastic, although large plastic items (> 5 mm) that may fragment into microplastic are addressed in the Municipal Regional Permit for Stormwater and the statewide trash amendments and requirements.

PRIORITY QUESTIONS

1 How much microplastic pollution is there in the Bay?

2 What are the health risks?

3 What are the sources, pathways, loadings, and processes leading to microplastic pollution in the Bay?

4 Have the concentrations of microplastic in the Bay increased or decreased?

5 What management actions may be effective in reducing microplastic pollution?

RECENT FINDINGS

Microplastics have been monitored in Bay surface water, sediment, prey fish, bivalves, and the adjacent ocean. Microplastics were ubiquitous, and the concentrations in Bay surface water were higher than other major water bodies monitored to date with comparable methods. Microplastics ingested by prey fish and bivalves were mostly fibers, and indicate microplastics are entering the Bay food web. Average concentrations of microplastics measured in Bay stormwater were 100 times greater than average concentrations in Bay wastewater effluent, emphasizing the importance of outdoor sources and emission of microplastics in urban areas.

Tire wear particles are a dominant source of microplastics that are transported to the Bay in urban runoff; this is supported by observations of tire wear particles in Bay Area stormwater samples, as well as literature estimates of 3-5 kg/yr of tire wear particles released per capita in the US. Fibers were the second most abundant type of particle observed in Bay stormwater samples, but the major source of the fibers is unclear. Based on ongoing literature synthesis to understand the sources of pathways of microplastics, our hypothesis is that cigarette filters and tumble-air dryers are important sources of fibers that warrant further investigation.

The Southern California Coastal Water Research Project, State Water Board, SFEI, and various academic experts have collaboratively developed an online database and web tool that summarizes published literature on microplastic effects on human and wildlife health. The tool, ToMEx: Toxicity of Microplastics Explorer, will be launched publicly this year and is proposed to be an online platform to share microplastic study results.

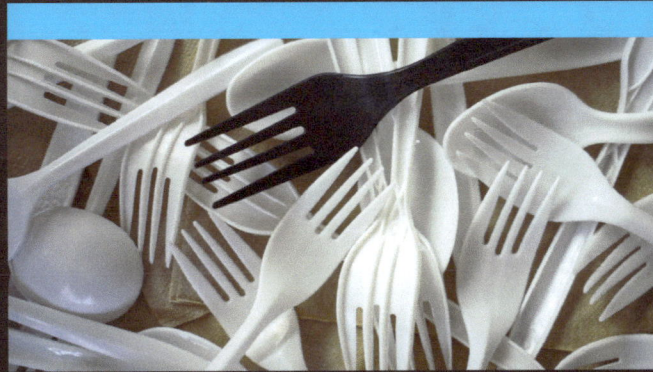

WORKPLAN HIGHLIGHTS

- Development of a conceptual model of microplastic sources and pathways to urban stormwater. Given the importance of urban stormwater as a pathway for microplastics in the environment, it is crucial to develop a conceptual understanding of the sources and sub-pathways for microplastics in urban stormwater to inform management actions that will reduce microplastics in San Francisco Bay. We are conducting a literature review and collecting data that will identify priorities for research and initial mitigation activities, providing support for Bay and statewide microplastic strategies and informing management efforts that will be effective in preventing microplastic pollution.

- Development of a tires science strategy to address stakeholder priorities. Evaluating water quality impacts of tire wear particles and tire wear chemical contaminants is a primary example of how investigations for the Microplastics Workgroup, Emerging Contaminants Workgroup, and Source Pathways and Loadings Workgroup continue to intersect. We will identify RMP priority data needs relating to tire contaminants that are not being addressed by others in the scientific community and develop a multi-year strategy to address those needs. Studies identified in this Tires Strategy could help recruit collaborators and funding sources beyond the RMP.

COLLABORATORS

- 5 Gyres Institute
- Bay Area Clean Water Agencies
- California Ocean Protection Council
- California State Water Resources Control Board
- Central Contra Costa Sanitary District
- City of Palo Alto
- East Bay Municipal Utility District
- Moss Landing Marine Laboratories
- Pacific Northwest Consortium on Plastics
- Patagonia
- University of Toronto

PROGRAM AREA
UPDATE | SEDIMENT

BACKGROUND

Sediment is critical to the health of the San Francisco Bay ecosystem. Suspended sediment concentrations in Bay water have an important role in controlling algae blooms and subsequent anoxia by limiting light availability. Sediment delivered to the Bay from the surrounding watersheds and transported within the Bay carries priority pollutants such as PCBs and mercury. Sediment deposition on tidal marshes and mudflats allows these habitats to increase in elevation and keep pace with rising sea level. Sediment is dredged from Bay shipping channels, harbors, and ports; some of this dredged sediment is removed from the Bay completely, which helps remove contaminants from the system, and some is beneficially reused in wetland restoration projects.

The RMP has been monitoring sediment in the Bay since the Program began in 1993. In recent years, sea level rise has heightened interest in sediment supply to the Bay. The mass balance and transport pathways of Bay sediment are critical factors controlling the degree to which mudflats, marshes, and other shoreline habitats get the sediment supply needed to be resilient over the long-term. As the San Francisco Bay Restoration Authority decides how to allocate $500 million over the next 20 years, it is critical to know the amount and quality of sediment available for restored tidal habitats.

In 2018, the RMP created a new Sediment Workgroup. The mission of the Workgroup is to provide technical oversight and stakeholder guidance on RMP studies addressing questions about sediment delivery, sediment transport, dredging, and beneficial reuse of sediment.

RELATION TO PERMIT REQUIREMENTS

Essential Fisheries Habitat Consultation, PCBs TMDL, Mercury TMDL

- Provides information for setting dredged material testing thresholds and in-Bay disposal limits

Long-Term Management Strategy for Dredged Material in San Francisco Bay

- Provides information about sediment mass balance in the whole Bay, submbayments, and margin areas

- Informs dredged sediment thresholds for beneficial reuse projects

USES OF PROGRAM AREA DATA FOR MANAGEMENT DECISIONS

- NOAA 2011 Programmatic Essential Fish Habitat Agreement and 2015 LTMS Amended Programmatic Biological Opinion

- Long-Term Management Strategy for Dredged Material in SF Bay (LTMS) to comply with the Basin Plan

- Regional Restoration Plans

- PCB TMDL

- Mercury TMDL

PRIORITY QUESTIONS

1. What are acceptable levels of chemicals in sediment for placement in the Bay, baylands, or restoration projects?

2. Are there effects on fish, benthic species, and submerged habitats from dredging or placement of sediment?

3. What are the sources, sinks, pathways, and loadings of sediment and sediment-bound contaminants to and within the Bay and submbayments?

4. How much sediment is passively reaching tidal marshes and restoration projects and how could the amounts be increased by management actions?

5. What are the concentrations of suspended sediment in the Estuary and its segments?

RECENT FINDINGS

A 2018 RMP special study used PCB data from the Dredged Material Management Office (DMMO) database to estimate PCB concentrations in dredged sediment, assess how dredged sediment concentrations compare to ambient concentrations, and assess the mass of PCBs moved to various disposal sites (in-Bay, deep ocean, and upland). PCB concentrations from sediment in dredged nearshore sites were found to be often similar to ambient RMP margin sites and higher than those in the ambient open-Bay and, but one to two orders of magnitude less than the most contaminated sites in the Bay. The study also found that approximately 50% of the PCB mass in dredged sediment is removed from the Bay via upland disposal and reuse.

Suspended sediment monitoring by USGS at Dumbarton Bridge in water year (WY) 2016 indicated that cohesive particle flocculation is an important factor for calculating the sediment flux into Lower South Bay. Based on these findings, the RMP allocated funds for additional studies at Dumbarton Bridge in Lower South Bay and at Benicia Bridge in the North Bay to investigate the importance of flocculation in sediment flux estimates. Analysis of suspended sediment flux at Dumbarton Bridge from WY2009-2016 indicated that when flocculation of cohesive sediment is accounted for, changes in the magnitude and direction in cumulative suspended sediment flux measurements were observed.

In WY2016 and WY2017, the USGS monitored the sediment flux through the Golden Gate. Results indicated net sediment flux into the Bay during a short period of high Delta and local tributary flow. Based on recommendations in the study report, the RMP funded a modeling study in 2020 that evaluated suspended sediment flux through the Golden Gate over the entire 2017 period of high Delta outflow and related findings from the model simulation to the USGS flux measurements. The model results showed net sediment flux out the Golden Gate, and that the duration of the flux analysis plays a large role in the estimated being in the flood direction (into the Bay) or ebb direction (out of the Bay).

WORKPLAN HIGHLIGHTS

- **Bathymetric change analysis.** Assessment of the changing bathymetry of the Bay is essential to understanding sediment transport and erosion dynamics. The RMP is funding the USGS to compile bathymetric data throughout the Bay and calculate bathymetric change since the 1990s. This work will show recent erosion and deposition dynamics at the subembayment scale and highlight the most pressing data gaps for assessing future Bay bathymetry.

- **Sediment delivery to marshes.** Salt marshes around the Bay provide critical habitat and natural shoreline protection. The RMP is funding the USGS to investigate the factors controlling sediment delivery to and deposition on a South Bay tidal marsh surface. The effort includes measurement of suspended sediment flux in the shallows adjacent to a marsh, flux into the marsh through a tidal creek, deposition and accretion on the marsh, and the variation in deposition with elevation and vegetation density and type. Results will be useful for prioritizing marsh restoration sites, assessing restoration actions, and understanding mechanisms of sediment delivery to and sea level rise vulnerability of marshes.

- **Bay sediment conceptual model.** This project will produce a detailed conceptual model of sediment dynamics for the Bay. The conceptual model will be incorporate recent monitoring and modeling results and working hypotheses of sediment dynamics within the Bay and between subembayments. It will also include an uncertainty analysis of key variables that affect sediment fate and transport at the subembayment scale for a range of time scales (e.g., tidal, seasonal, annual) for past and future conditions. The model will be used to inform policy decisions and build frameworks for management, monitoring, and numeric modeling.

COLLABORATORS

- Anchor QEA
- Bay Conservation and Development Commission
- Bay Planning Coalition
- Integral Consulting Inc.
- San Francisco Bay Regional Water Quality Control Board
- South Bay Salt Ponds Restoration Project
- US Army Corps of Engineers
- US Environmental Protection Agency
- US Geological Survey - Pacific Coastal and Marine Science Center
- US Geological Survey - Western Ecological Research Center

ACKNOWLEDGEMENTS

RMP STAFF

Kristin Art
Nina Buzby
Ariella Chelsky
Jay Davis
Scott Dusterhoff
Melissa Foley
Alicia Gilbreath
Cristina Grosso
Jennifer Hunt
Biruk Imagnu
Diana Lin
Jeremy Lowe
Lester McKee
Miguel Mendez
Ezra Miller
Kelly Moran
Derek Roberts
David Senn
Rebecca Sutton
Martin Trinh
Patrick Walsh
Michael Weaver
Adam Wong
Don Yee
Jamie Yin
Tan Zi

EDITORS

Jay Davis
Melissa Foley

CONTRIBUTING AUTHORS

Ariella Chelsky
Jay Davis
Scott Dusterhoff
Melissa Foley
Alicia Gilbreath
Diana Lin
Dave Senn
Rebecca Sutton
Don Yee

REVIEWERS

The following reviewers greatly improved this document by providing comments on draft versions:

Richard Looker
Luisa Valiela
Lester McKee
Kelly Moran

REPORT DESIGN

Ellen Plane
Ruth Askevold

ADDITIONAL PHOTO CREDITS

Melissa Foley – pages vi, 8-9, 40, 43 (upper), 44
Shira Bezalel – pages 25
Martin Trinh – pages 42, 43 (lower)
Alicia Gilbreath – page 46
Derek Roberts – page 49
Ariella Chelsky - page 50
Don Yee – pages 50, 51, 55
Diana Lin – page 52
Mark Eliot – page 53 (upper and lower)
Pete Kauhanen – page 54

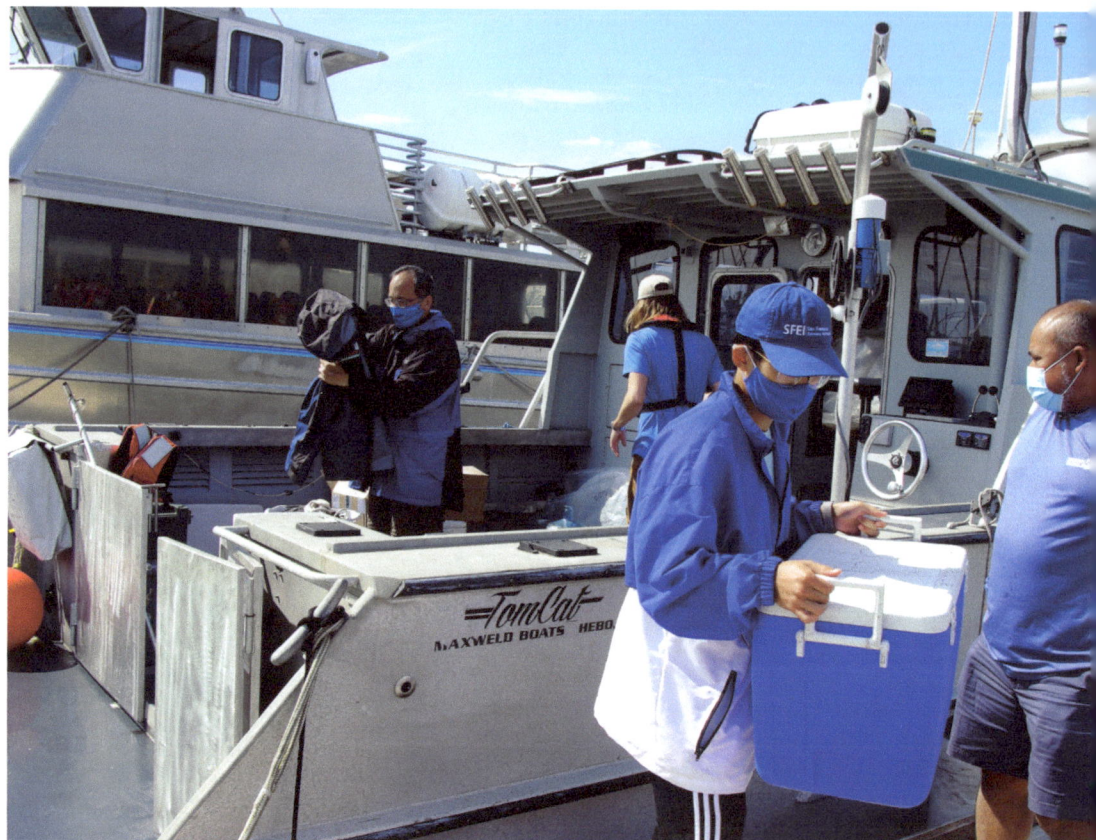

The end of the 2021 Water Cruise. Photograph by Melissa Foley.

RMP REGIONAL MONITORING PROGRAM FOR WATER QUALITY IN SAN FRANCISCO BAY

sfei.org/rmp

The RMP is administered by the San Francisco Estuary Institute
4911 Central Avenue, Richmond, CA 94804, p: 510-746-SFEI (7334)